CLEP* COLLEGE MATHEMATICS

Mel Friedman, M.S.

D1209554

Research & Education Association
Visit our website at: www.rea.com

WITHDRAWN

Research & Education Association
61 Ethel Road West
Piscataway, New Jersey 08854
E-mail: info@rea.com

CLEP College Mathematics

Library of Congress Control Number 2012938457

ISBN-13: 978-0-7386-1046-7
ISBN-10: 0-7386-1046-1

REA® is a registered trademark of
Research & Education Association, Inc.

CONTENTS

CHAPTER 4

CHAPTER 5

CHAPTER 6

CHAPTER 7

CHAPTER 8

ABOUT OUR AUTHOR

Mel Friedman, M.S., has a diversified background in mathematics, having taught both high school and college-level mathematics courses. Mr. Friedman was awarded his B.A. in Mathematics from Rutgers University and received his M.S. in Mathematics (with honors) from Fairleigh Dickinson University.

While pursuing his teaching career, Mr. Friedman also developed test items for Educational Testing Service and ACT, Inc. Formerly a professor at Kutztown University, Kutztown, Pa., Mr. Friedman is currently REA's Lead Mathematics Editor.

ABOUT RESEARCH & EDUCATION ASSOCIATION

Founded in 1959, Research & Education Association is dedicated to publishing the finest and most effective educational materials—including software, study guides, and test preps—for students in middle school, high school, college, graduate school, and beyond.

Today, REA's wide-ranging catalog is a leading resource for teachers, students, and professionals.

ACKNOWLEDGMENTS

We would like to thank Pam Weston, Publisher, for setting the quality standards for production integrity and managing the publication to completion; John Paul Cording, Vice President, Technology, for coordinating the design and development of the REA Study Center; Larry B. Kling, Vice President, Editorial, for his supervision of revisions and overall direction; Diane Goldschmidt and Michael Reynolds, Managing Editors, for coordinating development of this edition; Transcend Creative Services for typesetting this edition; and Weymouth Design and Christine Saul, Senior Graphic Designer, for designing our cover.

CHAPTER 1

Passing the CLEP College Mathematics Exam

PASSING THE CLEP COLLEGE MATHEMATICS EXAM

Congratulations! You're joining the millions of people who have discovered the value and educational advantage offered by the College Board's College-Level Examination Program, or CLEP. This test prep covers everything you need to know about the CLEP College Mathematics exam, and will help you earn the college credit you deserve while reducing your tuition costs.

GETTING STARTED

There are many different ways to prepare for a CLEP exam. What's best for you depends on how much time you have to study and how comfortable you are with the subject matter. To score your highest, you need a system that can be customized to fit you: your schedule, your learning style, and your current level of knowledge.

This book, and the online tools that come with it, allow you to create a personalized study plan through three simple steps: assessment of your knowledge, targeted review of exam content, and reinforcement in the areas where you need the most help.

Let's get started and see how this system works.

Test Yourself & Get Feedback	Score reports from your online diagnostic and practice tests give you a fast way to pinpoint what you already know and where you need to spend more time studying.
Review with the Book	Study the topics tested on the CLEP exam. Targeted review chapters cover everything you need to know.
Improve Your Score	Armed with your score reports, you can personalize your study plan. Review the parts of the book where you're weakest and study the answer explanations for the test questions you answered incorrectly.

THE REA STUDY CENTER

The best way to personalize your study plan and focus on your weaknesses is to get feedback on what you know and what you don't know. At the online REA Study Center, you can access two types of assessment: a diagnostic exam and full-length practice exams. Each of these tools provides true-to-format questions and delivers a detailed score report that follows the topics set by the College Board.

Diagnostic Exam

Before you begin your review with the book, take the online diagnostic exam. Use your score report to help evaluate your overall understanding of the subject, so you can focus your study on the topics where you need the most review.

Full-Length Practice Exams

These practice tests give you the most complete picture of your strengths and weaknesses. After you've finished reviewing with the book, test what you've learned by taking the first of the two online practice exams. Review your score report, then go back and study any topics you missed. Take the second practice test to ensure you have mastered the material and are ready for test day.

If you're studying and don't have Internet access, you can take the printed tests in the book. These are the same practice tests offered at the REA Study Center, but without the added benefits of timed testing conditions and diagnostic score reports. Because the actual exam is computer-based, we recommend you take at least one practice test online to simulate test-day conditions.

AN OVERVIEW OF THE EXAM

The CLEP College Mathematics exam consists of 60 multiple-choice questions, each with four possible answer choices, to be answered in 90 minutes.

The exam covers the material one would find in a college-level class for nonmathematics majors in fields not requiring knowledge of advanced mathematics. The exam places little emphasis on arithmetic and calculators are not allowed. A nongraphing calculator is provided to the test taker during the examination as part of the testing software.

The approximate breakdown of topics is as follows:

10%   Sets

10%   Logic

20%   Real Number Systems

20%   Functions and Their Graphs

25%   Probability and Statistics

15%   Additional Topics from Algebra and Geometry

ALL ABOUT THE CLEP PROGRAM

What is the CLEP?

CLEP is the most widely accepted credit-by-examination program in North America. CLEP exams are available in 33 subjects and test the material commonly required in an introductory-level college course. Examinees can earn from three to twelve credits at more than 2,900 colleges and universities in the

U.S. and Canada. For a complete list of the CLEP subject examinations offered, visit the College Board website: *www.collegeboard.org/clep*.

Who takes CLEP exams?

CLEP exams are typically taken by people who have acquired knowledge outside the classroom and who wish to bypass certain college courses and earn college credit. The CLEP program is designed to reward examinees for learning—no matter where or how that knowledge was acquired.

Although most CLEP examinees are adults returning to college, many graduating high school seniors, enrolled college students, military personnel, veterans, and international students take CLEP exams to earn college credit or to demonstrate their ability to perform at the college level. There are no prerequisites, such as age or educational status, for taking CLEP examinations. However, because policies on granting credits vary among colleges, you should contact the particular institution from which you wish to receive CLEP credit.

Who administers the exam?

CLEP exams are developed by the College Board, administered by Educational Testing Service (ETS), and involve the assistance of educators from throughout the United States. The test development process is designed and implemented to ensure that the content and difficulty level of the test are appropriate.

When and where is the exam given?

CLEP exams are administered year-round at more than 1,200 test centers in the United States and can be arranged for candidates abroad on request. To find the test center nearest you and to register for the exam, contact the CLEP Program:

CLEP Services
P.O. Box 6600
Princeton, NJ 08541-6600
Phone: (800) 257-9558 (8 A.M. to 6 P.M. ET)
Fax: (609) 771-7088
Website: *www.collegeboard.org/clep*

OPTIONS FOR MILITARY PERSONNEL AND VETERANS

CLEP exams are available free of charge to eligible military personnel and eligible civilian employees. All the CLEP exams are available at test centers on college campuses and military bases. Contact your Educational Services Officer or Navy College Education Specialist for more information. Visit the DANTES or College Board websites for details about CLEP opportunities for military personnel.

Eligible U.S. veterans can claim reimbursement for CLEP exams and administration fees pursuant to provisions of the Veterans Benefits Improvement Act of 2004. For details on eligibility and submitting a claim for reimbursement, visit the U.S. Department of Veterans Affairs website at *www.gibill.va.gov.*

CLEP can be used in conjunction with the Post-9/11 GI Bill, which applies to veterans returning from the Iraq and Afghanistan theaters of operation. Because the GI Bill provides tuition for up to 36 months, earning college credits with CLEP exams expedites academic progress and degree completion within the funded timeframe.

SSD ACCOMMODATIONS FOR CANDIDATES WITH DISABILITIES

Many test candidates qualify for extra time to take the CLEP exams, but you must make these arrangements in advance. For information, contact:

College Board Services for Students with Disabilities
P.O. Box 6226
Princeton, NJ 08541-6226
Phone: (609) 771-7137 (Monday through Friday, 8 A.M. to 6 P.M. ET)
TTY: (609) 882-4118
Fax: (609) 771-7944
E-mail: ssd@info.collegeboard.org

6-WEEK STUDY PLAN

Although our study plan is designed to be used in the six weeks before your exam, it can be condensed to three weeks by combining each two-week period into one.

Be sure to set aside enough time—at least two hours each day—to study. The more time you spend studying, the more prepared and relaxed you will feel on the day of the exam.

Week	Activity
1	Take the Diagnostic Exam. The score report will identify topics where you need the most review.
2–4	Study the review chapters. Use your diagnostic score report to focus your study.
5	Take Practice Test 1 at the REA Study Center. Review your score report and re-study any topics you missed.
6	Take Practice Test 2 at the REA Study Center to see how much your score has improved. If you still got a few questions wrong, go back to the review and study any topics you may have missed.

TEST-TAKING TIPS

Know the format of the test. CLEP computer-based tests are fixed-length tests. This makes them similar to the paper-and-pencil type of exam because you have the flexibility to go back and review your work in each section.

Learn the test structure, the time allotted for each section of the test, and the directions for each section. By learning this, you will know what is expected of you on test day, and you'll relieve your test anxiety.

Read all the questions—completely. Make sure you understand each question before looking for the right answer. Reread the question if it doesn't make sense.

Annotate the questions. Highlighting the key words in the questions will help you find the right answer choice.

Read all of the answers to a question. Just because you think you found the correct response right away, do not assume that it's the best answer. The last answer choice might be the correct answer.

Work quickly and steadily. You will have 90 minutes to answer 60 questions, so work quickly and steadily. Taking the timed practice tests online will help you learn how to budget your time.

Use the process of elimination. Stumped by a question? Don't make a random guess. Eliminate as many of the answer choices as possible. By eliminating just two answer choices, you give yourself a better chance of getting the item correct, since there will only be three choices left from which to make your guess. Remember, your score is based only on the number of questions you answer correctly.

Don't waste time! Don't spend too much time on any one question. Remember, your time is limited and pacing yourself is very important. Work on the easier questions first. Skip the difficult questions and go back to them if you have the time.

Look for clues to answers in other questions. If you skip a question you don't know the answer to, you might find a clue to the answer elsewhere on the test.

Acquaint yourself with the computer screen. Familiarize yourself with the CLEP computer screen beforehand by logging on to the College Board website. Waiting until test day to see what it looks like in the pretest tutorial risks injecting needless anxiety into your testing experience. Also, familiarizing yourself with the directions and format of the exam will save you valuable time on the day of the actual test.

Be sure that your answer registers before you go to the next item. Look at the screen to see that your mouse-click causes the pointer to darken the proper oval. If your answer doesn't register, you won't get credit for that question.

THE DAY OF THE EXAM

On test day, you should wake up early (after a good night's rest, of course) and have breakfast. Dress comfortably, so you are not distracted by being too hot or too cold while taking the test. (Note that "hoodies" are not allowed.) Arrive at the test center early. This will allow you to collect your thoughts and relax before the test, and it will also spare you the anxiety that comes with being late. As an added incentive, keep in mind that no one will be allowed into the test session after the test has begun.

Before you leave for the test center, make sure you have your admission form and another form of identification, which must contain a recent photograph, your name, and signature (i.e., driver's license, student identification card, or current alien registration card). You will not be admitted to the test center if you do not have proper identification.

You may wear a watch to the test center. However, you may not wear one that makes noise, because it may disturb the other test-takers. No cell phones, dictionaries, textbooks, notebooks, briefcases, or packages will be permitted, and drinking, smoking, and eating are prohibited.

Good luck on the CLEP College Mathematics exam!

CHAPTER 2

Sets

SETS

You will see the topics of set theory in most of the chapters of this book; in fact, you use set theory in many of your everyday activities. But since it is not labeled as "set theory" in most cases, you are unaware that set theory is the basis for most of your mathematical and logical thought. This chapter introduces the set theory vocabulary you should know as well as such topics as Venn diagrams for the union and intersection of sets (used in logic), laws of set operations (similar to those for operations on the real number system), and Cartesian products (used in graphs of linear functions). So let's set the stage for sets.

SETS

A **set** is defined as a collection of items. Each individual item belonging to a set is called an **element** or **member** of that set.

Sets are usually represented by capital letters, and elements by lowercase letters. If an item k belongs to a set A, we write $k \in A$ ("k is an element of A"). If k is not in A, we write $k \notin A$ ("k is not an element of A").

The order of the elements in a set does not matter:

$$\{1, 2, 3\} = \{3, 2, 1\} = \{1, 3, 2\}, \text{ etc.}$$

A set can be described in two ways:

1. element by element.

2. a rule characterizing the elements.

For example, given the set A of the whole numbers starting with 1 and ending with 9, we can describe it either as $A = \{1, 2, 3, 4, 5, 6, 7, 8, 9\}$ or as $A = \{$whole numbers greater than 0 and less than 10$\}$. In both methods, the description is enclosed in brackets. A kind of shorthand is often used for the second method of set description, so instead of writing out a complete sentence between the brackets, we can write instead

$$A = \{k \mid 0 < k < 10, k \text{ a whole number}\}$$

This is read as "the set of all elements k such that k is greater than 0 and less than 10, where k is a whole number."

A set not containing any members is called the **empty** or **null** set. It is written either as ϕ or $\{ \ \}$.

A set is **finite** if the number of its elements can be counted.

Example:

$\{2, 3, 4, 5\}$ is finite since it has four elements.

Example:

$\{3, 6, 9, 12, ..., 300\}$ is finite since it has 100 elements.

Note: The empty set, denoted by ϕ, is finite since we can count the number of elements it has, namely zero.

Any set that is not finite is called **infinite**.

Example:

$\{1, 2, 3, 4, ...\}$

Example:

$\{..., -7, -6, -5, -4\}$

Example:

$\{x \mid x \text{ is a real number between 4 and 5}\}$

SUBSETS

> Given two sets A and B, A is said to be a **subset** of B if every member of set A is also a member of set B.

A is a *proper* subset of B if B contains at least one element not in A. We write $A \subseteq B$ if A is a subset of B, and $A \subset B$ if A is a proper subset of B.

Two sets are **equal** if they have exactly the same elements; in addition, if $A = B$, then $A \subseteq B$ and $B \subseteq A$.

Example:

Let $A = \{1, 2, 3, 4, 5\}$

$B = \{1, 2\}$

$C = \{1, 4, 2, 3, 5\}$

1. A equals C, and A and C are subsets of each other, but not proper subsets.
2. $B \subseteq A$, $B \subseteq C$, $B \subset A$, $B \subset C$ (B is a subset of both A and C. In particular, B is a proper subset of A and C).

Two sets are **equivalent** if they have the same *number* of elements.

Example:

$O = \{3, 7, 9, 12\}$ and $E = \{4, 7, 12, 19\}$. O and E are equivalent sets, since each one has four elements.

Example:

$F = \{1, 3, 5, 7, ..., 99\}$ and $G = \{2, 4, 6, 8, ..., 100\}$. F and G are equivalent sets, since each one has 50 elements.

Note: If two sets are equal, they are automatically equivalent.

A **universal set** U is a set from which other sets draw their members. If A is a subset of U, then the complement of A, denoted A', is the set of all elements in the universal set that are not elements of A.

Example:

If $U = \{1, 2, 3, 4, 5, 6, ...\}$ and $A = \{1, 2, 3\}$, then $A' = \{4, 5, 6, ...\}$.

Figure 2-1 illustrates this concept through the use of a simple **Venn diagram**.

Figure 2-1

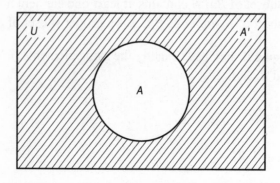

A Venn diagram is a visual way to show the relationships among or between sets that share something in common. Usually, the Venn diagram consists of two or more overlapping circles, with each circle representing a set of elements, or members. If two circles overlap, the members in the overlap belong to both sets; if three circles overlap, the members in the overlap belong to all three sets. Although Venn diagrams can be formed for any number of sets, you will probably encounter only two or three sets (circles) when working with Venn diagrams. As shown in Figure 1, the circles are usually drawn inside a rectangle called the universal set, which is the set of all possible members in the universe being described.

Venn diagrams are organizers. They are used to organize similarities (overlaps) and differences (non-overlaps of circles) visually, and they can pertain to any subject. For example, if the universe is all animals, Circle A may represent all animals that live in the water, and Circle B may represent all mammals. Then whales would be in the intersection of Circles A and B, but lobsters would be only in Circle A, humans would be only in Circle B, and scorpions would be in the part of the universe that was outside of Circles A and B. These relationships are shown in Figure 2-2.

Figure 2-2

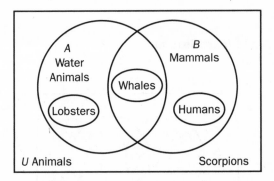

UNION AND INTERSECTION OF SETS

> The **union** of two sets A and B, denoted A ∪ B, is the set of all elements that are either in A or B or both.

Figure 2-3 is a Venn diagram for $A \cup B$. The shaded area represents the given operation.

Figure 2-3

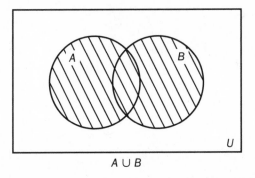

A ∪ B

> The **intersection** of two sets A and B, denoted A ∩ B, is the set of all elements that belong to both A and B.

Figure 2-4 is a Venn diagram for $A \cap B$. The shaded area represents the given operation.

If $A = \{1, 2, 3, 4, 5\}$ and $B = \{2, 3, 4, 5, 6\}$, then $A \cup B = \{1, 2, 3, 4, 5, 6\}$ and $A \cap B = \{2, 3, 4, 5\}$.

If $A \cap B = \phi, A$ and B are **disjoint**.

Figure 2-4

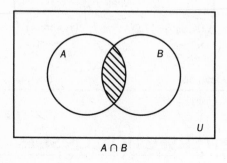

$A \cap B$

LAWS OF SET OPERATIONS

If U is the universal set, and A, B, C are any subsets of U, then the following hold for union, intersection, and complement:

Identity Laws

1a. $A \cup \phi = A$

1b. $A \cap \phi = \phi$

2a. $A \cup U = U$

2b. $A \cap U = A$

Idempotent Laws

3a. $A \cup A = A$

3b. $A \cap A = A$

Complement Laws

4a. $A \cup A' = U$

4b. $A \cap A' = \phi$

5a. $\phi' = U$

5b. $U' = \phi$

Commutative Laws

6a. $A \cup B = B \cup A$

6b. $A \cap B = B \cap A$

Associative Laws

7a. $(A \cup B) \cup C = A \cup (B \cup C)$

7b. $(A \cap B) \cap C = A \cap (B \cap C)$

Figures 2-5 and 2-6 illustrate the associative law for intersections. In Figure 2-5, the intersection of A and B is done first, and then the intersection of this result with C. In Figure 2-6, the intersection of B and C is done first, and then the intersection of this result with A. In both cases, the end result (double-hatched region) is the same.

Figure 2-5   **Figure 2-6**

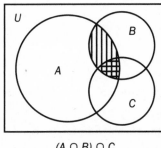   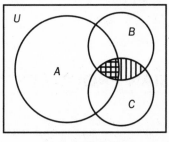

$(A \cap B) \cap C$ $A \cap (B \cap C)$

Distributive Laws

8a. $A \cup (B \cap C) = (A \cup B) \cap (A \cup C)$

8b. $A \cap (B \cup C) = (A \cap B) \cup (A \cap C)$

De Morgan's Laws

9a. $(A \cup B)' = A' \cap B'$

9b. $(A \cap B)' = A' \cup B'$

> The **difference** of two sets, A and B, written as $A - B$, is the set of all elements that belong to A but do not belong to B.

Example:

$J = \{10, 12, 14, 16\}, K = \{9, 10, 11, 12, 13\}$

$J - K = \{14, 16\}$. Note that $K - J = \{9, 11, 13\}$.

In general, $J - K \neq K - J$.

Example:

$T = \{a, b, c\}, V = \{a, b, c, d, e, f\}$.

$T - V = \phi$, whereas $V - T = \{d, e, f\}$.

Note, in this example, that T is a proper subset of V. In general, whenever set A is a proper subset of set B, $A - B = \phi$.

If set P is any set, then $P - \phi = P$ and $\phi - P = \phi$. Also if P and Q are any sets, $P - Q = P \cap Q'$.

CARTESIAN PRODUCT

Given two sets M and N, the **Cartesian product**, denoted $M \times N$, is the set of all ordered pairs of elements in which the first component is a member of M and the second component is a member of N.

Often, the elements of the Cartesian product can be found by making a table with the elements of the first set as row headings and the elements of the second set as column headings, and the elements of the table the pairs formed from these elements.

Example:

$M = \{1, 3, 5\}, N = \{2, 8\}$
The Cartesian product $M \times N = \{(1, 2), (1, 8), (3, 2), (3, 8), (5, 2), (5, 8)\}$

We can easily see that these are all of the elements of $M \times N$ and the only elements of $M \times N$ by looking at Table 2-1:

Table 2-1 $M \times N$

	2	8
1	1, 2	1, 8
3	3, 2	3,8
5	5, 2	5, 8

Example:

$W = \{a, b, c\}$, $Y = \{a, g, h\}$
The Cartesian product $W \times Y = \{(a, a), (a, g), (a, h), (b, a), (b, g), (b, h), (c, a), (c, g), (c, h)\}$

The elements for this Cartesian product are shown in Table 2-2.

Table 2-2 $W \times Y$

	a	g	h
a	a, a	a, g	a, h
b	b, a	b, g	b, h
c	c, a	c, g	c, h

In the first example above, since M has 3 elements and N has 2 elements, $M \times N$ has $3 \times 2 = 6$ elements. In the second example above, since W has 3 elements and Y has 3 elements, $W \times Y$ has $3 \times 3 = 9$ elements. In general, if the first set has x elements and the second set has y elements, the Cartesian product will have xy elements.

Drill Questions

1. Suppose that set K has 12 elements and set L has 3 elements. How many elements are there in the Cartesian product $K \times L$?

    (A) 4
    (B) 9
    (C) 15
    (D) 36

2. If set M has 10 elements and set N has 7 elements, which one of the following statements must be true?

    (A) The maximum number of elements in M' is 3.
    (B) The minimum number of elements in $M \cap N$ is 7.
    (C) The maximum number of elements in $M \cup N$ is 17.
    (D) The minimum number of elements in $M - N$ is 10.

3. If $A = \{x \mid x$ is an even integer less than 10$\}$ and $B = \{$all negative numbers$\}$, which one of the following describes $A \cap B$?

    (A) {all negative numbers and all positive even integers}
    (B) {all negative numbers}
    (C) {all negative even integers}
    (D) {all negative odd integers}

4. Consider the Venn diagram shown below. Which one of the following correctly describes the shading?

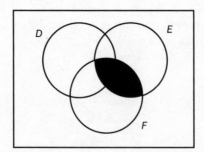

    (A) $E \cap F$
    (B) $D \cap E \cap F$
    (C) $D \cap E$
    (D) $D \cap F$

5. Which one of the following is an example of disjoint sets?

   (A) $\{0, 1, 2, 3\}$ and $\{3, 2, 1, 0\}$
   (B) $\{0, 2, 4, 6\}$ and $\{2, 4, 6, 8\}$
   (C) $\{0, 3, 6, 9\}$ and $\{9, 16, 25, 36\}$
   (D) $\{0, 4, 8, 12\}$ and $\{6, 10, 14, 18\}$

6. If the universal set $U = \{x \mid x$ is a positive odd integer less than 30$\}$, $R = \{1, 5, 7\}$, and $S = \{1, 3, 7, 11, 13\}$, how many elements are in $(R \cap S)'$?

   (A) 15
   (B) 13
   (C) 7
   (D) 2

7. If $P \subseteq Q$, which one of the following conclusions must be true?

   (A) P is either equal to Q or P is a proper subset of Q.
   (B) P is a proper subset of Q.
   (C) Q is a proper subset of P.
   (D) P is either equal to Q or P is the empty set.

8. Given any two sets F and G, which one of the following is not necessarily true?

   (A) $F \cup G = G \cup F$
   (B) $F \cap G = G \cap F$
   (C) $F - G = G - F$
   (D) $F \cap F' = \phi$

9. Which one of the following is a finite set?

   (A) $\{x \mid x$ is an irrational number less than 0$\}$
   (B) $\{x \mid x$ is a positive rational number less than 20$\}$
   (C) $\{x \mid x$ is a negative fraction greater than $-8\}$
   (D) $\{x \mid x$ is a natural number less than 50$\}$

10. Consider the Venn diagram shown below.

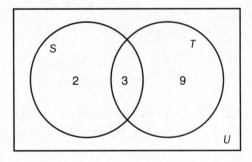

If the given numbers represent the number of elements in each region, how many elements are in $T - S$?

(A) 10

(B) 9

(C) 7

(D) 6

Answers to Drill Questions

1. **(D)** Given two sets K and L, the Cartesian product $K \times L$ is the set consisting of all ordered pairs in which the first element is chosen from K and the second element is chosen from L. For example, if set K contains the element a and set L contains the element b, then one of the elements of $K \times L$ is (a, b). Since we know that K has 12 elements and L has 3 elements, we conclude that there are $(12)(3) = 36$ different ordered pairs for $K \times L$.

2. **(C)** $M \cup N$ is the union of sets M and N, and it represents all the elements that belong to at least one of M or N. The maximum number of elements occurs when M and N have no common elements. In this case, the number of elements in $M \cup N$ becomes the sum of the number of elements in each of M and N, which is $10 + 7 = 17$.

3. **(C)** In roster form, $A = \{..., -8, -6, -4, -2, 0, 2, 4, 6, 8\}$. Although we cannot write set B in roster form, we can determine that the elements common to both A and B can be represented as $\{..., -8, -6, -4, -2\}$, which is the set of all negative even integers.

4. **(A)** The shaded region represents all elements common to both sets E and F. This is the definition of the intersection of sets E and F, written as $E \cap F$. Note that the presence of set D becomes incidental, and does not affect the correct answer choice.

5. **(D)** Disjoint sets are those that do not contain any common elements, such as $\{0, 4, 8, 12\}$ and $\{6, 10, 14, 18\}$.

6. **(B)** Written in roster form, $U = \{1, 3, 5, 7, \ldots, 29\}$, which contains 15 elements. $R \cap S$ represents the elements common to both R and S, so $R \cap S = \{1, 7\}$. Finally, $(R \cap S)'$ represents the elements in U that do not belong to $R \cap S$. Thus, $(R \cap S)'$ must contain $15 - 2 = 13$ elements.

7. **(A)** $P \subseteq Q$ means that each element of set P is also an element of set Q. This implies that either (a) P and Q are identical, or (b) Q contains all the elements of P, plus at least one element not found in P. By definition, if part (b) applies, then P is a proper subset of Q.

8. **(C)** $F - G$ represents the set of elements in F that do not belong to G, whereas $G - F$ represents the set of elements in G that do not belong to F. Unless these sets are equivalent, $F - G \neq G - F$. For example, let $F = \{1, 3, 6\}$ and let $G = \{1, 5, 8, 9\}$. Then $F - G = \{3, 6\}$, but $G - F = \{5, 8, 9\}$.

9. **(D)** A finite set is one in which the number of elements can be counted. The set of natural numbers less than 50 can be written in roster form as $\{1, 2, 3, 4, \ldots, 49\}$, which has 49 elements.

10. **(B)** $T - S$ is the set of elements in T that do not belong to S. According to the given Venn Diagram, there are 9 elements that fit this description.

CHAPTER 3

The Real Number System

THE REAL NUMBER SYSTEM

Real numbers provide the basis for most precalculus mathematics topics.

> **Real numbers** are all of the numbers on the **number line** (see Figure 3-1).

In fact, a nice way to visualize real numbers is that they can be put in a one-to-one correspondence with the set of all points on a line. Real numbers include positives, negatives, square roots, π (pi), and just about any number you have ever encountered.

Figure 3-1

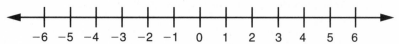

PROPERTIES OF REAL NUMBERS

Similar to the Laws of Set Operations presented in the last chapter, real numbers have several properties that you should know. You have used some of these properties ever since you could count. You intuitively knew that $3 + 2$ gives the same result as $2 + 3$, or that if you add 0 to a number it remains unchanged. These properties deal with addition and multiplication. They do not work for subtraction and division. For example, you also intuitively know that $3 - 2$ is not the same as $2 - 3$.

Perhaps you are not familiar with the names of these properties. The following list provides the names for these properties. Learn them—you will encounter these property names in the next chapter and on mathematics tests. The examples use the numbers 2, 3, and 4, but the rules apply to any real numbers.

COMMUTATIVE PROPERTY

The numbers *commute*, or move:

Addition $\qquad$ $2 + 3 = 3 + 2$

Multiplication $\qquad$ $2 \times 3 = 3 \times 2$

ASSOCIATIVE PROPERTY

The numbers can be grouped, or *associated*, in any order:

Addition $\qquad$ $2 + (3 + 4) = (2 + 3) + 4$

Multiplication $\qquad$ $2 \times (3 \times 4) = (2 \times 3) \times 4$

Later in this chapter, you will see that operations in parentheses are always done first, but the associative property says that you can move the parentheses and it won't make a difference.

DISTRIBUTIVE PROPERTY

The first number gets *distributed* to the ones in parentheses:

$$2 \times (3 + 4) = (2 \times 3) + (2 \times 4)$$

The following properties have to do with the special numbers 0 and 1:

IDENTITY PROPERTY

Adding 0 or multiplying by 1 doesn't change the original value:

Addition $\qquad$ $3 + 0 = 3$

Multiplication $\qquad$ $3 \times 1 = 3$

INVERSE PROPERTY

The *inverse* of addition is subtraction and the *inverse* of multiplication is division:

Additive Inverse $\qquad$ $3 + (-3) = 0$

Multiplicative Inverse $\qquad$ $3 \times \dfrac{1}{3} = 1$

Note that the multiplicative inverse doesn't work for 0 because division by 0 is not defined.

COMPONENTS OF REAL NUMBERS

The set of all real numbers (designated as R) has various components:

$N = \{1, 2, 3, \ldots\}$, the set of all **natural numbers**

$W = \{0, 1, 2, 3, \ldots\}$, the set of all **whole numbers**

$I = \{\ldots, -3, -2, -1, 0, 1, 2, 3, \ldots\}$, the set of all **integers**

$Q = \left\{ \frac{a}{b} \mid a, b \in I \text{ and } b \neq 0 \right\}$, the set of all **rational numbers**

$S = \{x \mid x$ has a decimal that is nonterminating and does not have a repeating block$\}$, the set of all **irrational numbers**.

It is obvious that $N \subseteq W$, $W \subseteq I$, and $I \subseteq Q$, but a similar relationship does not hold between Q and S. More specifically, the decimal names for elements of Q are either (1) terminating or (2) nonterminating with a repeating block.

Examples of rational numbers include $\frac{1}{2} = .5$ and $\frac{1}{3} = .333\ldots$

This means that Q and S have no common elements.

Examples of irrational numbers include $.101001000\ldots$, π, and $\sqrt{2}$.

All real numbers are normally represented by R and $R = Q \cup S$. This means that every real number is either rational or irrational.

FRACTIONS

All rational numbers can be displayed as **fractions**, which consist of a numerator (on the top) and a denominator (on the bottom).

> **Proper fractions** are numbers between -1 and $+1$; the numerator is less than the denominator. Examples of proper fractions are $\frac{1}{2}$, $\frac{3}{4}$, and $\frac{17}{19}$.
>
> **Improper fractions** are all other rational numbers; the numerator is greater than or equal to the denominator.

Improper fractions are also called mixed numbers because they can be written as a whole number with a fractional part. Examples of improper fractions are $\frac{2}{1}$, $\frac{4}{3}$, and $\frac{19}{17}$. The first of these is actually a whole number (2); the others are equivalent to the mixed numbers $1\frac{1}{3}$ and $1\frac{2}{17}$, respectively.

ODD AND EVEN NUMBERS

When dealing with odd and even numbers, keep in mind the following:

Adding:

even + even = even

odd + odd = even

even + odd = odd

Multiplying:

even × even = even

even × odd = even

odd × odd = odd

FACTORS AND DIVISIBILITY NUMBERS

> Any counting number that divides into another number with no remainder is called a **factor** of that number.

The factors of 20 are 1, 2, 4, 5, 10, and 20.

> Any number that can be divided by another number with no remainder is called a **multiple** of that number.

Examples of multiples of 20 are 20, 40, 60, 80, etc.

PROBLEM

List the factors and multiples of 28.

SOLUTION

The factors of 28 are 1, 2, 4, 7, 14, and 28. Some multiples of 28 are 28, 56, 84, and 112. Note that the list of multiples is endless.

ABSOLUTE VALUE

The **absolute value** of a number is represented by two vertical lines around the number and is equal to the positive value, regardless of sign.

The absolute value of a real number A is defined as follows:

$$|A| = \begin{cases} A \text{ if } A \geq 0 \\ -A \text{ if } A < 0 \end{cases}$$

Examples:

$$|5| = 5$$
$$|-8| = -(-8) = 8$$

Absolute values follow the given rules:

1. $|-A| = |A|$

2. $|A| \geq 0$, equality holding only if $A = 0$

3. $\left|\dfrac{A}{B}\right| = \dfrac{|A|}{|B|}, B \neq 0$

4. $|AB| = |A| \times |B|$

5. $|A|^2 = A^2$

Calculate the value of each of the following expressions:

1. $||2 - 5| + 6 - 14|$
2. $|-5| \times |4| + \dfrac{|-12|}{4}$

SOLUTION

Before solving this problem, one must remember the order of operations: parentheses, multiplication and division, addition and subtraction.

1. $||-3| + 6 - 14| = |3 + 6 - 14| = |9 - 14| = |-5| = 5$
2. $(5 \times 4) + \dfrac{12}{4} = 20 + 3 = 23$

INTEGERS

There are various subsets of I, the set of all integers:

NEGATIVE INTEGERS

The set of integers starting with -1 and decreasing:

$$\{-1, -2, -3, \ldots\}.$$

EVEN INTEGERS

The set of integers divisible by 2:

$$\{\ldots, -4, -2, 0, 2, 4, 6, \ldots\}.$$

ODD INTEGERS

The set of integers not divisible by 2:

$$\{\ldots, -3, -1, 1, 3, 5, 7, \ldots\}.$$

CONSECUTIVE INTEGERS

The set of integers that differ by 1:

$$\{n, n + 1, n + 2, \ldots\} \ (n = \text{an integer}).$$

PRIME NUMBERS

The set of positive integers greater than 1 that are divisible only by 1 and themselves:

$$\{2, 3, 5, 7, 11, \ldots\}.$$

COMPOSITE NUMBERS

The set of integers, other than 0 and ± 1, that are not prime.

PROBLEM

Classify each of the following numbers into as many different sets as possible.

Example: real, integer ...

1. 0   3. $\sqrt{6}$ 5. $\dfrac{2}{3}$ 7. 11

2. 9   4. $\dfrac{1}{2}$ 6. 1.5

SOLUTION

1. 0 is a real number, an integer, a whole number, and a rational number.

2. 9 is a real number, an odd number, a natural number, and a rational number.

3. $\sqrt{6}$ is a real number, and an irrational number.

4. $\dfrac{1}{2}$ is a real number, and a rational number.

5. $\dfrac{2}{3}$ is a real number, and a rational number.

6. 1.5 is a real number, a decimal, and a rational number.

7. 11 is a prime number, an odd number, a real number, a natural number, and a rational number.

INEQUALITIES

If x and y are real numbers, then one and only one of the following statements is true:

$$x > y, x = y, \text{ or } x < y.$$

This is the **order property of real numbers**.

If a, b, and c are real numbers, the following statements are true:

If $a < b$ and $b < c$, then $a < c$.

If $a > b$ and $b > c$, then $a > c$.

This is the **transitive property of inequalities**.

If a, b, and c are real numbers and $a > b$, then $a + c > b + c$ and $a - c > b - c$. This is the **addition property of inequality**.

> An **inequality** is a statement in which the value of one quantity or expression is greater than ($>$), less than ($<$), greater than or equal to ($\geq$), less than or equal to ($\leq$), or not equal to ($\neq$) that of another.

Example:

$5 > 4$. This expression means that the value of 5 is greater than the value of 4.

The **graph of an inequality** in one variable is represented by either a ray or a line segment on the real number line.

The endpoint is not a solution if the variable is strictly less than or greater than a particular value. In those cases, the endpoint is indicated by an open circle.

Example:

$x > 2$

2 is not a solution and should be represented as shown.

The endpoint is a solution if the variable is either (1) less than or equal to or (2) greater than or equal to a particular value. In those cases, the endpoint is indicated by a closed circle.

Example:

$5 > x \geq 2$

In this case, 2 is a solution and 5 is not a solution, and the solution should be represented as shown.

Example:

$x < 2$ or $x > 5$

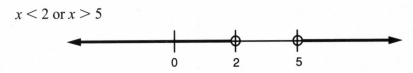

In this case, neither 2 nor 5 is a solution. Thus, an open circle must be shown at $x = 2$ and at $x = 5$.

Example:

$x \leq 2$ and $x \geq 5$

In this case, there is no solution. It is impossible for a number to be both no greater than 2 and no less than 5.

Example:

$x \geq 2$ or $x \leq 5$

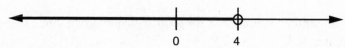

In this case, the solution is all real numbers. Any number *must* belong to at least one of these inequalities. Some numbers, such as 3, belong to both inequalities. If you graph these inequalities separately, you will notice two rays going in opposite directions and which overlap between 2 and 5, inclusive.

Intervals on the number line represent sets of points that satisfy the conditions of an inequality.

An **open** interval does not include any endpoints.

Example:

$\{x \mid x > -3\}$, read as "the set of values x such that $x > -3$." The graph would appear as:

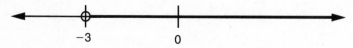

Example:

$\{x \mid x < 4\}$. The graph would appear as:

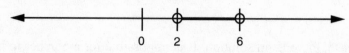

Example:

$\{x \mid 2 < x < 6\}$. The graph would appear as:

Example:

$\{x \mid x$ is any real number$\}$. The graph would appear as:

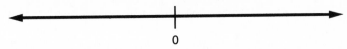

A **closed** interval includes two endpoints.

Example:

$\{x \mid -5 \le x \le 2\}$. The graph would appear as:

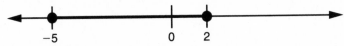

A **half-open** interval includes one endpoint.

Example:

$\{x \mid x \ge 3\}$. The graph would appear as:

Example:

$\{x \mid x \le 6\}$. The graph would appear as:

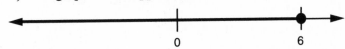

Example:

$\{x \mid -4 < x \le -1\}$.

Example:

$\{x \mid -2 \le x < 1\}$.

Drill Questions

1. If c is any odd integer, which one of the following must be an even integer?

    (A) $\dfrac{c}{2}$
    (B) $3c - 1$
    (C) $c^2 + 2$
    (D) $2c + 1$

2. What is the value of $|\,7 - 13\,| - |\,-2 - 9\,|$?

    (A) 16
    (B) 5
    (C) -5
    (D) -16

3. Which one of the following is an irrational number whose value lies between 7 and 8?

    (A) $\sqrt[3]{400}$
    (B) $7.\overline{2}$
    (C) $\sqrt{65}$
    (D) $\dfrac{25}{3}$

4. Which one of the following numbers is a prime number between 90 and 100?

    (A) 91
    (B) 94
    (C) 95
    (D) 97

5. Which one of the following inequalities describes a graph on the number line that includes all real numbers?

    (A) $x \geq 1$ and $x \leq 4$
    (B) $x \geq 1$ or $x \leq 4$
    (C) $x \leq 1$ or $x \geq 4$
    (D) $x \leq 1$ and $x \geq 4$

6. If 15 is a factor of x, which one of the following is true for any x?

   (A) 30 must be a factor of x.
   (B) Each of 3 and 5 must be factors of x.
   (C) x must be a prime number greater than 15.
   (D) The only prime factors of x are 3 and 5.

7. The number $0.\overline{4}$ is equivalent to which fraction?

   (A) $\dfrac{4}{5}$

   (B) $\dfrac{4}{7}$

   (C) $\dfrac{4}{9}$

   (D) $\dfrac{4}{11}$

8. Not including the number 16, how many factors of 16 are there?

   (A) 7
   (B) 6
   (C) 5
   (D) 4

9. n^2 is a proper fraction in reduced form. Which one of the following must be true?

   (A) n is a proper fraction.
   (B) n is an improper fraction.
   (C) n is a positive number.
   (D) n is a negative number.

10. Which one of the following has no solution for x?

   (A) $|x| > 0$
   (B) $|x| = 0$
   (C) $|x| = -x$
   (D) $|x| < 0$

Answers to Drill Questions

1. **(B)** The product of 3 and an odd integer must be an odd integer. The difference of an odd integer and 1 must be an even integer. For example, suppose $c = 5$. Then $3c - 1 = (3)(5) - 1 = 15 - 1 = 14$, which is an even integer.

2. **(C)** Recall that the absolute value of any quantity must be nonnegative. Then $|7 - 13| = |-6| = 6$ and $|-2 - 9| = |-11| = 11$. Then $6 - 11 = 6 + (-11) = -5$.

3. **(A)** $\sqrt[3]{400}$ is irrational because we cannot find any integer whose cube is exactly 400. Note that since $7^3 = 343$ and $8^3 = 512$, we know that $\sqrt[3]{400}$ has a value between 7 and 8. (It is approximately 7.368.)

4. **(D)** A prime number can be divided by only two numbers, itself and 1. The number 97 is prime because it is divisble by only 1 and 97.

5. **(B)** The inequality $x \geq 1$ includes all numbers greater than or equal to 1. The inequality $x \leq 4$ includes all numbers less than or equal to 4. Every number must satisfy at least one of these conditions. (Some numbers, such as 2, satisfy both conditions.) Thus, the graph of $x \geq 1$ or $x \leq 4$ includes all numbers on the number line.

6. **(B)** If 15 is a factor of x, then x must be divisible by 15 and by any factor of 15. Each of 3 and 5 is a factor of 15. Let $x = 30$. Note that 15 is a factor of 30, since $30 \div 15 = 2$. In addition, we note that each of 3 and 5 is a factor of 30, since $30 \div 3 = 10$ and $30 \div 5 = 6$.

7. **(C)** Let $N = 0.\overline{4}$, so that $10N = 4.\overline{4}$. By subtracting the first equation from the second equation, we get $9N = 4$. Thus, $N = \frac{4}{9}$. Note that we can check this answer by long division. (Divide 9 into 4 to get $0.\overline{4}$.)

8. **(D)** Not including the number 16, the other four factors of 16 are 1, 2, 4, and 8.

9. **(A)** If n^2 is a proper fraction in reduced form, then its numerator and denominator must contain no common factors. This implies that its square root, namely n, also contains no common factors. So n must be proper. Note that the numerator and denominator of n^2 must be a perfect square. As an example, let $n^2 = \frac{81}{100}$. Then $n = \sqrt{\frac{81}{100}} = \frac{9}{10}$, which is a proper fraction.

10. **(D)** The absolute value of any number must be nonnegative. This means that $|x|$ must be greater than or equal to zero. Thus, $|x| < 0$ has no solution.

CHAPTER 4

Algebra Topics

CHAPTER 4

ALGEBRA TOPICS

This chapter discusses the algebra topics that you should know to score well on your CLEP exam, such as exponents, logarithms, complex numbers, and inequalities. In addition, it presents applications from algebra so you can see the usefulness of this subject.

EXPONENTS

When a number is multiplied by itself a specific number of times, it is said to be **raised to a power**. The way this is written is $a^n = b$, where a is the number or **base**; n is the **exponent** or **power** that indicates the number of times the base appears when multiplied by itself; and b is the **product** of this multiplication.

In the expression 3^2, 3 is the base and 2 is the exponent. This means that 3 appears 2 times when multiplied by itself (3×3), and the product is 9.

An exponent can be either positive or negative. A negative exponent implies a fraction, such that if n is a negative integer

$$a^{-n} = \frac{1}{a^n}, a \neq 0$$

So, $2^{-4} = \frac{1}{2^4} = \frac{1}{16}$.

The **reciprocal** of a number is 1 divided by that number. The exception is 0, since $\frac{1}{0}$ is undefined.

We see that a negative exponent yields a fraction that is the reciprocal of the original base and exponent, with the exponent now positive instead of negative. Essentially, any quantity raised to a negative exponent can be "flipped" to the other side of the fraction bar and its exponent changed to a positive exponent. For example,

$$4^{-3} = \frac{1}{4^3} = \frac{1}{64}$$

An exponent that is 0 gives a result of 1, assuming that the base itself is not equal to 0.

$$a^0 = 1, a \neq 0$$

An exponent can also be a fraction. If m and n are positive integers,

$$a^{\frac{m}{n}} = \sqrt[n]{a^m}$$

The numerator remains the exponent of a, but the denominator tells what root to take. For example,

$$4^{\frac{3}{2}} = \sqrt[2]{4^3} = \sqrt{64} = 8$$
$$3^{\frac{4}{2}} = \sqrt[2]{3^4} = \sqrt{81} = 9$$

If a fractional exponent was negative, the operation involves the reciprocal as well as the roots. For example,

$$27^{-\frac{2}{3}} = \frac{1}{27^{\frac{2}{3}}} = \frac{1}{\sqrt[3]{27^2}} = \frac{1}{\sqrt[3]{729}} = \frac{1}{9}$$

PROBLEMS

Simplify the following expressions:

1. -3^{-2}

2. $(-3)^{-2}$

3. $\dfrac{-3}{4^{-1}}$

4. $-16^{-\frac{1}{2}}$

SOLUTIONS

1. Here the exponent applies only to 3. Since

$$x^{-y} = \frac{1}{x^y}, \text{ so } -3^{-2} = -(3)^{-2} = -\left(\frac{1}{3^2}\right) = -\frac{1}{9}$$

2. In this case, the exponent applies to the negative base. Thus,

$$(-3)^{-2} = \frac{1}{(-3)^2} = \frac{1}{(-3)(-3)} = \frac{1}{9}$$

3. $$\frac{-3}{4^{-1}} = \frac{-3}{\left(\frac{1}{4}\right)^1} = \frac{-3}{\frac{1^1}{4^1}} = \frac{-3}{\frac{1}{4}} = \frac{-3}{1} \times \frac{4}{1} = -12$$

4. $$-16^{-\frac{1}{2}} = -\frac{1}{16^{\frac{1}{2}}} = -\frac{1}{\sqrt{16}} = -\frac{1}{4}$$

GENERAL LAWS OF EXPONENTS

$a^p a^q = a^{p+q}$, bases must be the same

$$4^2 4^3 = 4^{2+3} = 4^5 = 1{,}024$$

$(a^p)^q = a^{pq}$

$$(2^3)^2 = 2^6 = 64$$

$\dfrac{a^p}{a^q} = a^{p-q}$, bases must be the same, $a \neq 0$

$$\frac{3^6}{3^2} = 3^{6-2} = 3^4 = 81$$

$(ab)^p = a^p b^p$

$$(3 \times 2)^2 = 3^2 \times 2^2 = (9)(4) = 36$$

$\left(\dfrac{a}{b}\right)^p = \dfrac{a^p}{b^p}, b \neq 0$

$$\left(\frac{4}{5}\right)^2 = \frac{4^2}{5^2} = \frac{16}{25}$$

LOGARITHMS

An equation

$$y = b^x$$

(with $b > 0$ and $b \neq 1$) is called an **exponential function**.

The exponential functions with base b can be written as

$$y = f(x) = b^x.$$

The inverse of an exponential function is the **logarithmic function**,

$$f^{-1}(x) = \log_b x.$$

Inverse is denoted by $f^{-1}(\ )$; this doesn't mean a negative exponent.

PROBLEM

Write the following equations in logarithmic form:

$3^4 = 81$ and $M^k = 5$.

SOLUTION

The expression $y = b^x$ is equivalent to the logarithmic expression $\log_b y = x$. Therefore, $3^4 = 81$ is equivalent to the logarithmic expression, $\log_3 81 = 4$.

$M^k = 5$ is equivalent to the logarithmic expression $\log_M 5 = k$.

PROBLEM

Find the value of x for $\log_5 25 = x$ and $\log_4 x = 2$.

SOLUTION

$\log_5 25 = x$ is equivalent to $5^x = 25$. Thus $x = 2$, since $5^2 = 25$, and $\log_5 25 = 2$.

$\log_4 x = 2$ is equivalent to $4^2 = x$, or $x = 16$.

LOGARITHM PROPERTIES

If M, N, p, and b are positive numbers and $b \neq 1$, then

$\log_b 1 = 0$

$\log_b b = 1$

$\log_b b^x = x$

$\log_b (MN) = \log_b M + \log_b N$

$\log_b (M/N) = \log_b M - \log_b N$

$\log_b M^p = p \log_b M$

PROBLEM

If $\log_{10} 3 = .4771$ and $\log_{10} 4 = .6021$, find $\log_{10} 12$.

SOLUTION

Since $12 = 4(3)$, $\log_{10} 12 = \log_{10} (4) (3)$.

Remember, $\log_b(MN) = \log_b M + \log_b N$.

Therefore, $\log_{10} 12 = \log_{10} 4 + \log_{10} 3 = .6021 + .4771 = 1.0792$

EQUATIONS

An **equation** is defined as a statement that two separate expressions are equal. A **solution** to an equation containing a single variable is a number that makes the equation true when it is substituted for the variable.

For example, in the equation $3x = 18$, 6 is the solution since $3(6) = 18$. Depending on the equation, there can be more than one solution. Equations with the same solutions are said to be **equivalent equations**. An equation without a solution is said to have a solution set that is the **empty** or **null** set, represented by ϕ.

Replacing an expression within an equation by an equivalent expression will result in a new equation with solutions equivalent to the original equation. For example, suppose we are given the equation

$$3x + y + x + 2y = 15.$$

By combining like terms, we get

$$3x + y + x + 2y = 4x + 3y.$$

Since these two expressions are equivalent, we can substitute the simpler form into the equation to get

$$4x + 3y = 15$$

Performing the same operation to both sides of an equation by the same expression will result in a new equation that is equivalent to the original equation.

ADDITION OR SUBTRACTION

$$y + 6 = 10$$

We can add (-6) to both sides:

$$y + 6 + (-6) = 10 + (-6)$$

$$y + 0 = 10 - 6 = 4$$

MULTIPLICATION OR DIVISION

$$3x = 6$$

We can divide both sides by 3:

$$\frac{3x}{3} = \frac{6}{3}$$

$$x = 2$$

So $3x = 6$ is equivalent to $x = 2$.

PROBLEM

Solve for x, justifying each step.

$$3x - 8 = 7x + 8$$

SOLUTION

$3x - 8 = 7x + 8$

$3x - 8 + 8 = 7x + 8 + 8$ Add 8 to both sides

$3x + 0 = 7x + 16$ Additive inverse property

$3x = 7x + 16$ Additive identity property

$3x - 7x = 7x + 16 - 7x$ Add $(-7x)$ to both sides

$-4x = 7x - 7x + 16$ Commutative property

$-4x = 0 + 16$ Additive inverse property

$-4x = 16$ Additive identity property

$\dfrac{-4x}{-4} = \dfrac{16}{-4}$ Divide both sides by -4

$x = -4$

CHECK YOUR WORK!

Replacing x with -4 in the original equation:

$$3x - 8 = 7x + 8$$

$$3(-4) - 8 = 7(-4) + 8$$

$$-12 - 8 = -28 + 8$$

$$-20 = -20$$

LINEAR EQUATIONS

> A **linear equation** with one unknown is one that can be put into the form $ax + b = 0$, where a and b are constants, and $a \neq 0$.

To solve a linear equation means to transform it into the form $x = \dfrac{-b}{a}$.

A. If the equation has unknowns on both sides of the equality, it is convenient to put similar terms on the same sides. Refer to the following example:

$4x + 3 = 2x + 9$

$4x + 3 - 2x = 2x + 9 - 2x$ Add $-2x$ to both sides

$(4x - 2x) + 3 = (2x - 2x) + 9$ Commutative property

$2x + 3 = 0 + 9$ Additive inverse property

$2x + 3 - 3 = 0 + 9 - 3$ Add -3 to both sides

$2x = 6$ Additive inverse property

$\dfrac{2x}{2} = \dfrac{6}{2}$ Divide both sides by 2

$x = 3$

B. If the equation appears in fractional form, it is necessary to transform it using cross-multiplication, and then repeat the same procedure as in (A). For example,

$$\frac{3x + 4}{3} = \frac{7x + 2}{5}$$

Cross-multiply as follows:

$$\frac{3x + 4}{3} \;\;\diagdown\!\!\!\!\diagup\;\; \frac{7x + 2}{5}$$

to obtain:

$$3(7x + 2) = 5(3x + 4).$$

This is equivalent to:

$$21x + 6 = 15x + 20,$$

which can be solved as in (A).

$21x + 6 = 15x + 20$

$21x - 15x + 6 = 15x - 15x + 20$ Add $-15x$ to both sides

$6x + 6 - 6 = 20 - 6$ Combine like terms and add -6 to both sides

$6x = 14$ Combine like terms

$\dfrac{6x}{6} = \dfrac{14}{6}$ Divide both sides by 6

$x = \dfrac{7}{3}$

FACTOR THEOREM

If $x = c$ is a solution of the equation $f(x) = 0$, then $(x - c)$ is a **factor** of $f(x)$.

Example:

Let $f(x) = 2x^2 - 5x - 3$. By inspection, we can determine that $2(3)^2 - (5)(3) - 3 = (2)(9) - (5)(3) - 3 = 0$. In this example, $c = 3$, so $(x - 3)$ is also a factor of $2x^2 - 5x - 3$.

Example:

Let $f(x) = x^3 + 3x^2 - 4$. By inspection, we can determine that $(-2)^3 + 3(-2)^2 - 4 = -8 + (3)(4) - 4 = 0$. In this example, $c = -2$, so $(x + 2)$ is also a factor of $x^3 + 3x^2 - 4$.

REMAINDER THEOREM

> If a is any constant and if the polynomial $p(x)$ is divided by $(x - a)$, the **remainder** is $p(a)$.

Example:

Given a polynomial $p(x) = 2x^3 - x^2 + x + 4$, divided by $x - 1$, the remainder is $P(1) = 2(1)^3 - (1)^2 + 1 + 4 = 6$.

That is, $2x^3 - x^2 + x + 4 = q(x) + \dfrac{6}{(x - 1)}$, where $q(x)$ is a polynomial.

Note that in this case $a = 1$.

Also, by using long division, we get $q(x) = 2x^2 + x + 2$.

Example:

Given a polynomial $p(x) = x^4 + x - 50$ divided by $x + 3$, the remainder is $p(-3) = 28$.

That is, $x^4 + x - 50 = q(x) + \dfrac{28}{(x + 3)}$, where $q(x)$ is a polynomial.

Note that in this case $a = -3$.

Also, by using long division, we get $q(x) = x^3 - 3x^2 + 9x - 26$.

SIMULTANEOUS LINEAR EQUATIONS

Two or more equations of the form $ax + by = c$, where a, b, c are constants and a, $b \neq 0$ are called **linear equations** with two unknown variables, or **simultaneous equations**.

Equations with more than one unknown variable are solvable only if you have as many equations as unknown variables.

There are several ways to solve systems of linear equations with two variables. Three of the basic methods are:

Method 1: **Substitution**—Find the value of one unknown in terms of the other. Substitute this value in the other equation and solve.

Method 2: **Addition or subtraction**—If necessary, multiply the equations by numbers that will make the coefficients of one unknown in the resulting equations numerically equal. If the signs of equal coefficients are the same, subtract the equations, otherwise add. The result is one equation with one unknown; we solve it and substitute the value into the other equations to find the unknown that we first eliminated.

Method 3: **Graph**—Graph both equations. The point of intersection of the drawn lines is a simultaneous solution for the equations, and its coordinates correspond to the answer that would be found by substitution or addition/ subtraction.

> A system of linear equations is **consistent** if there is only one solution for the system.
>
> A system of linear equations is **inconsistent** if it does not have any solutions.

Inconsistent equations represent parallel lines, which are discussed later in this chapter.

PROBLEM

Solve the system of equations.

1. $x + y = 3$
2. $3x - 2y = 14$

SOLUTION

Method 1 (Substitution): From equation (1), we get $y = 3 - x$. Substitute this value into equation (2) to get

$$3x - 2(3 - x) = 14$$
$$3x - 6 + 2x = 14$$
$$5x = 20$$
$$x = 4$$

Substitute $x = 4$ into either of the original equations to find $y = -1$.

The answer is $x = 4, y = -1$.

Method 2 (Addition or subtraction): If we multiply equation (1) by 2 and add the result to equation (2), we get

$$2x + 2y = 6$$
$$+ \ 3x - 2y = 14$$
$$\overline{5x + 0 = 20}$$
$$x = 4$$

Then, as in Method 1, substitute $x = 4$ into either of the original equations to find $y = -1$.

The answer is $x = 4, y = -1$.

Method 3 (Graphing): Find the point of intersection of the graphs of the equations. To graph these linear equations, solve for y in terms of x. The equations will be in the form $y = mx + b$, where m is the slope and b is the intercept on the y-axis. This is the **slope-intercept form** of the equation.

$$x + y = 3$$

Subtract x from both sides:

$$y = 3 - x$$

$$3x - 2y = 14$$

Subtract $3x$ from both sides:

$$-2y = 14 - 3x$$

Divide by -2:

$$y = -7 + \frac{3}{2}x$$

The graph of each of the linear functions can be determined by plotting only two points. For example, for $y = 3 - x$, let $x = 0$, then $y = 3$. Let $x = 1$, then $y = 2$. The two points on this first line are $(0, 3)$ and $(1, 2)$. For $y = -7 + \frac{3}{2}x$, let $x = 0$, then $y = -7$. Let $x = 2$, then $y = -4$. The two points on this second line are $(0, -7)$ and $(2, -4)$.

To find the point of intersection P of the two lines, graph them.

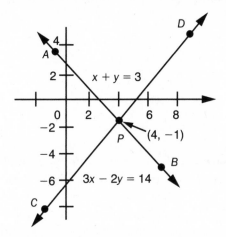

AB is the graph of equation (1), and CD is the graph of equation (2). The point of intersection P of the two graphs is the only point on both lines. The coordinates of P satisfy both equations and represent the desired solution of the problem. From the graph, P seems to be point $(4, -1)$. These coordinates satisfy both equations, and hence are the exact coordinates of the point of intersection of the two lines.

CHECK YOUR WORK!

To show that $(4, -1)$ satisfies both equations, substitute this point into both equations.

$x + y = 3$	$3x - 2y = 14$
$4 + (-1) = 3$	$3(4) - 2(-1) = 14$
$4 - 1 = 3$	$12 + 2 = 14$
$3 = 3$	$14 = 14$

DEPENDENT EQUATIONS

> **Dependent equations** are equations that represent the same line; therefore, every point on the line of a dependent equation represents a solution.

Since there are an infinite number of points on a line, there are an infinite number of simultaneous solutions.

Example:

$$2x + y = 8$$
$$4x + 2y = 16$$

These equations are dependent. Since they represent the same line, all points that satisfy either of the equations are solutions of the system.

PROBLEM

Solve the equations $2x + 3y = 6$ and $y = -\left(\dfrac{2x}{3}\right) + 2$ simultaneously.

SOLUTION

We have two equations and two unknowns:

$$2x + 3y = 6$$

and

$$y = -\left(\frac{2x}{3}\right) + 2$$

As with all simultaneous equations, there are several methods of solution. Since equation (2) already gives us an expression for y, we use the method of substitution.

Substitute $-\left(\dfrac{2x}{3}\right) + 2$ for y in equation (1):

$$2x + 3\left(-\frac{2x}{3} + 2\right) = 6$$

Distribute:

$$2x - 2x + 6 = 6$$

$$6 = 6$$

Although the result $6 = 6$ is true, it indicates no single solution for x. No matter what real number x is, if y is determined by equation (1), then equation (1) will always be satisfied.

The reason for this peculiarity may be seen if we take a closer look at the equation $y = -\left(\dfrac{2x}{3}\right) + 2$. It is equivalent to $3y = -2x + 6$, or $2x + 3y = 6$.

In other words, the two equations are equivalent. Any pair of values of x and y that satisfies one satisfies the other.

It is hardly necessary to verify that in this case the graphs of the given equations are identical lines, and that there are an infinite number of simultaneous solutions to these equations.

PARALLEL LINES

> Given two linear equations in x, y, their graphs are **parallel** lines if their slopes are equal. If the lines are parallel, they have no simultaneous solution.

Example:

Line $l_1 : 2x - 7y = 14$, Line $l_2 : 2x - 7y = 56$

In the slope-intercept form, the equation for l_1 is $y = \dfrac{2}{7}x - 2$ and the equation for l_2 is $y = \dfrac{2}{7}x - 8$. Each line has a slope of $\dfrac{2}{7}$.

PROBLEM

Solve the equations $2x + 3y = 6$ and $4x + 6y = 7$ simultaneously.

SOLUTION

We have two equations and two unknowns:

$$2x + 3y = 6$$

and

$$4x + 6y = 7$$

Again, there are several methods to solve this problem. We have chosen to multiply each equation by a different number so that when the two equations are added, one of the variables drops out. Thus,

Multiply equation (1) by 2: $\qquad 4x + 6y = 12 \qquad (3)$

Multiply equation (2) by -1: $\qquad \underline{+ \;\; -4x - 6y = -7} \qquad (4)$

Add equations (3) and (4): $\qquad\qquad\qquad 0 = 5$

We obtain a peculiar result!

Actually, what we have shown in this case is that there is no simultaneous solution to the given equations because $0 \neq 5$. Therefore, there is no simultaneous solution to these two equations, and hence no point satisfying both.

The straight lines that are the graphs of these equations must be parallel if they never intersect, but not identical, which can be seen from the graph of these equations.

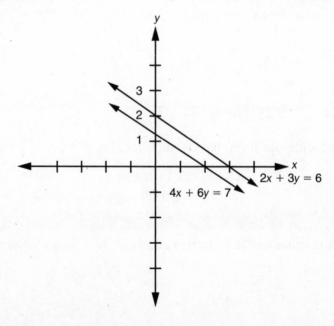

PERPENDICULAR LINES

> If the slopes of the graphs of two lines are negative reciprocals of each other, the lines are **perpendicular** to each other.

An example of two numbers that are negative reciprocals of each other are 2 and $-\frac{1}{2}$. (Remember: $2 = \frac{2}{1}$.)

Example:

l_3: $5x + 6y = 30$, l_4: $6x - 5y = 90$

In the slope-intercept form, the equation for l_3 is $y = -\frac{5}{6}x + 5$ and the equation for l_4 is $y = \frac{6}{5}x - 18$. The slope of l_3, which is $-\frac{5}{6}$, is the negative reciprocal of the slope of l_4, which is $\frac{6}{5}$. Therefore, l_3 is perpendicular to l_4.

To summarize:

- Parallel lines have slopes that are equal.
- Perpendicular lines have slopes that are negative reciprocals of each other.

ABSOLUTE VALUE EQUATIONS

> The **absolute value** of a, denoted $|a|$, is defined as
>
> $|a| = a$ when $a > 0$,
>
> $|a| = -a$ when $a < 0$,
>
> $|a| = 0$ when $a = 0$.

When the definition of absolute value is applied to an equation, the quantity within the absolute value symbol may have two values. This value can be either positive or negative before the absolute value is taken. As a result, each absolute value equation actually contains two separate equations.

When evaluating equations containing absolute values, proceed as follows:

Example:

$|5 - 3x| = 7$ is valid if either

$$5 - 3x = 7 \quad \text{or} \quad 5 - 3x = -7$$
$$-3x = 2 \qquad\qquad -3x = -12$$
$$x = -\frac{2}{3} \qquad\qquad x = 4$$

The solution set is therefore $x = \left(-\dfrac{2}{3}, 4\right)$

Remember, the absolute value of a number cannot be negative. So the equation $|5x + 4| = -3$ would have no solution.

INEQUALITIES

The solution of a given inequality in one variable x consists of all values of x for which the inequality is true.

> A **conditional inequality** is an inequality whose validity depends on the values of the variables in the sentence. That is, certain values of the variables will make the sentence true, and others will make it false.

The sentence $3 - y > 3 + y$ is a conditional inequality for the set of real numbers, since it is true for any replacement less than 0 and false for all others, or $y < 0$ is the solution set.

An **absolute inequality** for the set of real numbers means that for *any* real value for the variable, x, the sentence is always true.

The sentence $x + 5 > x + 2$ is an absolute inequality because the expression on the left is greater than the expression on the right.

A sentence is **inconsistent** if it is always false when its variables assume allowable values.

The sentence $x + 10 < x + 5$ is inconsistent because the expression on the left side is always greater than the expression on the right side.

The sentence $5y < 2y + y$ is inconsistent for the set of non-negative real numbers. For any y greater than 0, the sentence is always false.

Two inequalities are said to have the same **sense** if their signs of inequality point in the same direction.

The sense of an inequality remains the same if both sides are multiplied or divided by the same *positive* real number.

Example:

For the inequality $4 > 3$, if we multiply both sides by 5, we will obtain:

$$4 \times 5 > 3 \times 5$$

$$20 > 15$$

The sense of the inequality does not change.

If each side of an inequality is multiplied or divided by the same *negative* real number, however, the sense of an inequality becomes opposite.

Example:

For the inequality $4 > 3$, if we multiply both sides by -5, we would obtain:

$$4 \times (-5) < 3 \times (-5)$$
$$-20 < -15.$$

The sense of the inequality becomes opposite.

If $a > b$ and a, b, and n are positive real numbers, then

$$a^n > b^n \text{ and } a^{-n} < b^{-n}$$

If $x > y$ and $q > p$, then $x + q > y + p$.

If $x > y > 0$ and $q > p > 0$, then $xq > yp$.

> Inequalities that have the same solution set are called **equivalent inequalities**.

PROBLEM

Solve the inequality $2x + 5 > 9$.

SOLUTION

$$2x + 5 > 9$$

Add -5 to both sides: $\quad 2x + 5 + (-5) > 9 + (-5)$

Additive inverse property: $\quad 2x + 0 > 9 + (-5)$

Additive identity property: $\quad 2x > 9 + (-5)$

Combine terms: $\quad 2x > 4$

Multiply both sides by $\frac{1}{2}$ (this is the same as dividing both sides by 2): $\quad \frac{1}{2}(2x) > \frac{1}{2} \times 4$

$$x > 2$$

The solution set is

$$X = \{x \mid x > 2\}$$

(that is, all x, such that x is greater than 2).

COMPLEX NUMBERS

As indicated above, real numbers provide the basis for most precalculus mathematics topics. However, on occasion, real numbers by themselves are not enough to explain what is happening. As a result, complex numbers were developed.

> A **complex number** is a number that can be written in the form $a + bi$, where a and b are real numbers and $i = \sqrt{-1}$. The number a is the **real part**, and the number bi is the **imaginary part** of the complex number.

Returning momentarily to real numbers, the square of a real number cannot be negative. More specifically, the square of a positive real number is positive, the square of a negative real number is positive, and the square of 0 is 0.

i is defined to be a number with a property that $i^2 = -1$. Obviously, i is not a real number. C is then used to represent the set of all complex numbers:

$$C = \{a + bi \mid a \text{ and } b \text{ are real numbers}\}.$$

ADDITION, SUBTRACTION, AND MULTIPLICATION OF COMPLEX NUMBERS

Here are the definitions of addition, subtraction, and multiplication of complex numbers.

Suppose $x + yi$ and $z + wi$ are complex numbers. Then (remembering that $i^2 = -1$):

$$(x + yi) + (z + wi) = (x + z) + (y + w)i$$
$$(x + yi) - (z + wi) = (x - z) + (y - w)i$$
$$(x + yi) \times (z + wi) = (xz - wy) + (xw + yz)i$$

PROBLEM

Simplify $(3 + i)(2 + i)$.

$$(3 + i)(2 + i) = 3(2 + i) + i(2 + i)$$
$$= 6 + 3i + 2i + i^2$$
$$= 6 + (3 + 2)i + (-1)$$
$$= 5 + 5i$$

DIVISION OF COMPLEX NUMBERS

Division of two complex numbers is usually accomplished with a special procedure that involves the conjugate of a complex number. The conjugate of $a + bi$ is denoted by $\overline{a + bi}$ and defined by $\overline{a + bi} = a - bi$.

Also,

$$(a + bi)(a - bi) = a^2 + b^2$$

The usual procedure for division is to multiply and divide by the conjugate as shown below. Remember that multiplication and division by the same quantity leaves the original expression unchanged.

$$\frac{x + yi}{z + wi} = \frac{x + yi}{z + wi} \times \frac{z - wi}{z - wi}$$

$$= \frac{(xz + yw) + (-xw + yz)i}{z^2 + w^2}$$

$$= \frac{xz + yw}{z^2 + w^2} + \frac{-xw + yz}{z^2 + w^2}i$$

If a is a real number, then a can be expressed in the form $a = a + 0i$. Hence, every real number is a complex number and $R \subseteq C$.

All the properties of real numbers described in Chapter 2 carry over to complex numbers, so those properties will not be stated again.

QUADRATIC EQUATIONS

Consider the polynomial:

$ax^2 + bx + c = 0$, where $a \neq 0$.

This type of equation is called a **quadratic equation**.

There are several methods to solve quadratic equations, some of which are highlighted here. The first two are based on the fact that if the product of two factors is 0, either one or the other of the factors equals 0. The equation can be solved by setting each factor equal to 0. If you cannot see the factors right away, however, the quadratic formula, which is the last method presented, *always* works.

SOLUTION BY FACTORING

We are looking for two binomials that, when multiplied together, give the quadratic trinomial

$$ax^2 + bx + c = 0$$

This method works easily if $a = 1$ and you can find two numbers whose product equals c and sum equals b. The signs of b and c need to be considered:

- If c is positive, the factors are going to both have the same sign, which is b's sign.

- If c is negative, the factors are going to have opposite signs, with the larger factor having b's sign.

Once you have the two factors, insert them in the general factor format $(x + \_)$ $(x + \_) = 0$.

To solve the quadratic equation, set each factor equal to 0 to yield the solution set for x.

PROBLEM

Solve the quadratic equation $x^2 + 7x + 12 = 0$.

SOLUTION

We need two numbers whose product is $+12$ and sum is $+7$. They would be $+3$ and $+4$, and the quadratic equation would factor to $(x + 3)(x + 4) = 0$.

Therefore, $x + 3 = 0$ or $x + 4 = 0$ would yield the solutions, which are $x = -3, x = -4$.

CHECK YOUR WORK!

Substitute the values into the original quadratic equation:

For $x = -3, (-3)^2 + 7(-3) + 12 = 0$, or $9 + (-21) + 12 = 0$. So $x = -3$ is a solution.

Likewise, for $x = -4, (-4)^2 + 7(-4) + 12 = 0$, or $16 + (-28) + 12 = 0$. So $x = -4$ is a solution.

PROBLEM

Suppose the quadratic equation is similar to the previous example, but the sign of b is negative. Solve the quadratic equation $x^2 - 7x + 12 = 0$.

SOLUTION

We need two numbers whose product is $+12$ and sum is -7. They would be -3 and -4, and the quadratic would factor to $(x - 3)(x - 4) = 0$.

Therefore, $x - 3 = 0$ or $x - 4 = 0$ would yield the solutions, which are $x = 3, x = 4$.

CHECK YOUR WORK!

Substitute the values into the original quadratic equation:

For $x = 3, (3)^2 - 7(3) + 12 = 0$, or $9 - 21 + 12 = 0$. So $x = 3$ is a solution.

Likewise, for $x = 4, (4)^2 - 7(4) + 12 = 0$, or $16 - 28 + 12 = 0$. So $x = 4$ is a solution.

PROBLEM

As a final example, solve the quadratic equation $x^2 + 4x - 12 = 0$.

SOLUTION

We need two numbers whose product is -12 and sum is $+4$. They would be $+6$ and -2 (note that the larger numeral gets the $+$ sign, the sign of b). The quadratic equation would factor to $(x + 6)(x - 2) = 0$.

Therefore, $x + 6 = 0$ or $x - 2 = 0$ would yield the solutions, which are $x = -6, x = 2$.

CHECK YOUR WORK!

Substitute the values into the original quadratic equation:

For $x = -6$, $(-6)^2 + 4(-6) - 12 = 0$, or $36 + (-24) - 12 = 0$. So $x = -6$ is a solution.

Likewise, for $x = 2$, $(2)^2 + 4(2) - 12 = 0$, or $4 + 8 - 12 = 0$. So $x = 2$ is a solution.

SUM OF TWO SQUARES

If the quadratic consists of the difference of only two terms of the form $ax^2 - c$, and you can recognize them as perfect squares, the factors are simply the sum and difference of the square roots of the two terms. Note that a is a perfect square, but not necessarily 1, for this method.

PROBLEM

Solve the quadratic equation $x^2 - 16 = 0$.

SOLUTION

This is the difference of two perfect squares, x^2 and 16, whose square roots are x and 4. So the factors are $(x + 4)(x - 4) = 0$, and the solution is $x = \pm 4$.

For $x = 4$, $(4)^2 - 16 = 16 - 16 = 0$.

For $x = -4$, $(-4)^2 - 16 = 16 - 16 = 0$.

PROBLEM

Solve the quadratic equation $9x^2 - 36 = 0$.

SOLUTION

This is the difference of two perfect squares, $9x^2$ and 36, whose square roots are $3x$ and 6. So the factors are $(3x + 6)(3x - 6) = 0$, then $3x + 6 = 0$ or $3x - 6 = 0$, and the solution is $x = \pm 2$.

CHECK YOUR WORK!

For $x = 2$, $9(2)^2 - 36 = 36 - 36 = 0$.

For $x = -2$, $9(-2)^2 - 36 = 36 - 36 = 0$.

QUADRATIC FORMULA

If the quadratic equation does not have obvious factors, the roots of the equation can always be determined by the **quadratic formula** in terms of the coefficients a, b, and c as shown below:

$$x = \frac{-b \pm \sqrt{b^2 - 4ac}}{2a}$$

where $(b^2 - 4ac)$ is called the **discriminant** of the quadratic equation.

- If the discriminant is less than zero ($b^2 - 4ac < 0$), the roots are complex numbers, since the discriminant appears under a radical and square roots of negatives are imaginary numbers. A real number added to an imaginary number yields a complex number.

- If the discriminant is equal to zero ($b^2 - 4ac = 0$), the roots are real and equal.

- If the discriminant is greater than zero ($b^2 - 4ac > 0$), then the roots are real and unequal. The roots are rational if and only if a and b are rational and ($b^2 - 4ac$) is a perfect square; otherwise, the roots are irrational.

PROBLEMS

Compute the value of the discriminant and then determine the nature of the roots of each of the following four equations:

1. $4x^2 - 12x + 9 = 0$

2. $3x^2 - 7x - 6 = 0$

3. $5x^2 + 2x - 9 = 0$

4. $x^2 + 3x + 5 = 0$

SOLUTIONS

1. $4x^2 - 12x + 9 = 0$

   Here, a, b, and c are integers:

   $a = 4$, $b = -12$, and $c = 9$.

   Therefore,

   $b^2 - 4ac = (-12)^2 - 4(4)(9) = 144 - 144 = 0$

   Since the discriminant is 0, the roots are rational and equal.

2. $3x^2 - 7x - 6 = 0$

   Here, a, b, and c are integers:

   $a = 3$, $b = -7$, and $c = -6$.

   Therefore,

   $b^2 - 4ac = (-7)^2 - 4(3)(-6) = 49 + 72 = 121 = 11^2$.

   Since the discriminant is a perfect square, the roots are rational and unequal.

3. $5x^2 + 2x - 9 = 0$

    Here, a, b, and c are integers:

    $a = 5$, $b = 2$, and $c = -9$

    Therefore,

    $b^2 - 4ac = 2^2 - 4(5)(-9) = 4 + 180 = 184.$

    Since the discriminant is greater than zero, but not a perfect square, the roots are irrational and unequal.

4. $x^2 + 3x + 5 = 0$

    Here, a, b, and c are integers:

    $a = 1$, $b = 3$, and $c = 5$

    Therefore,

    $b^2 - 4ac = 3^2 - 4(1)(5) = 9 - 20 = -11$

    Since the discriminant is negative, the roots are complex.

PROBLEM

Solve the equation $x^2 - x + 1 = 0$.

SOLUTION

In this equation, $a = 1$, $b = -1$, and $c = 1$. Substitute into the quadratic formula.

$$x = \frac{-(-1) \pm \sqrt{(-1)^2 - 4(1)(1)}}{2(1)}$$

$$= \frac{1 \pm \sqrt{1 - 4}}{2}$$

$$= \frac{1 \pm \sqrt{-3}}{2}$$

$$= \frac{1 \pm \sqrt{3}i}{2}$$

$$x = \frac{1 + \sqrt{3}i}{2} \text{ or } x = \frac{1 - \sqrt{3}i}{2}$$

ADVANCED ALGEBRAIC THEOREMS

A. Every polynomial equation $f(x) = 0$ of degree greater than zero has at least one root either real or complex. This is known as the **fundamental theorem of algebra**.

B. Every polynomial equation of degree n has exactly n roots.

C. If a polynomial equation $f(x) = 0$ with real coefficients has a root $a + bi$, then the conjugate of this complex number $a - bi$ is also a root of $f(x) = 0$.

D. If $a + \sqrt{b}$ is a root of polynomial equation $f(x) = 0$ with rational coefficients, then $a - \sqrt{b}$ is also a root, where a and b are rational and $\sqrt{b}$ is irrational.

E. If a rational fraction in lowest terms $\dfrac{b}{c}$ is a root of the equation $a_n x^n + a_{n-1} x^{n-1} + \ldots + a_1 x + a_0 = 0$, $a_0 \neq 0$, and the a_i are integers, then b is a factor of a_0 and c is a factor of a_n.

F. Any rational roots of the equation $x^n + q_1 x^{n-1} + q_2 x^{n-2} + \ldots + q_{n-1} x + q_n = 0$ must be integers and factors of q_n. Note that $q_1, q_2, \ldots, q_n$ are integers.

Drill Questions

1. Which one of the following is an equation of a line parallel to the graph of $5x + 9y = 14$?

    (A) $y = \dfrac{5}{9}x + 14$

    (B) $y = -\dfrac{5}{9}x - 14$

    (C) $y = -\dfrac{9}{5}x + 14$

    (D) $y = -\dfrac{9}{5}x - 14$

2. The solution set of the inequality $5 - 7x \geq -9$ is

    (A) $X = \{x \mid x \leq 2\}$

    (B) $X = \{x \mid x \geq 2\}$

    (C) $X = \left\{x \mid x \leq \dfrac{4}{7}\right\}$

    (D) $X = \left\{x \mid x \geq \dfrac{4}{7}\right\}$

3. What is the value of $\left(\dfrac{-3}{4^{-1}}\right)(-3)^{-2}$?

    (A) $-\dfrac{4}{3}$

    (B) $-\dfrac{1}{9}$

    (C) $\dfrac{1}{9}$

    (D) $\dfrac{4}{3}$

4. If $\log_{10} P = 0.5$ and $\log_{10} Q^2 = 1.2$, what is the value of $\log_{10} PQ$?

    (A) 1.9

    (B) 1.7

    (C) 1.1

    (D) 0.6

5. What is the solution set for x in the following inequality?
$$-6 < \frac{2}{3}x + 6 < 2$$

(A) $-12 < x < 0$
(B) $-6 < x < 0$
(C) $-18 < x < -12$
(D) $-18 < x < -6$

6. What is the simplified expression for $-5i^7 + (2i)(i^2-2)$?

(A) $11i$
(B) $-11i$
(C) i
(D) $-i$

7. If $x + 2$ is a factor of $3x^3 - x + k$, what is the value of k?

(A) 22
(B) 16
(C) -16
(D) -22

8. A piece of lumber that is 28 inches in length is divided into three pieces. The length of the first piece is x inches and the length of the second piece is four times the length of the first piece. Which expression represents the length of the third piece?

(A) $28 - 4x$
(B) $28 - 5x$
(C) $4x - 28$
(D) $5x - 28$

9. What is the value of y in the following system of equations?

   $4x + 2y = 5$

   $3x + 5y = 9$

   (A) $\dfrac{1}{2}$

   (B) $\dfrac{3}{2}$

   (C) $\dfrac{5}{2}$

   (D) $\dfrac{7}{2}$

10. $2^5 = 32$ is equivalent to which logarithmic expression?
    (A) $\log_5 32 = 2$
    (B) $\log_{32} 5 = 2$
    (C) $\log_2 32 = 5$
    (D) $\log_2 5 = 32$

Answers to Drill Questions

1. **(B)** Rewrite the equation $5x + 9y = 14$ in slope-intercept form. Then $9y = -5x + 14$, which leads to $y = -\dfrac{5}{9}x + \dfrac{14}{9}$, where $-\dfrac{5}{9}$ is the slope and $\dfrac{14}{9}$ is the y-intercept. The equation of any line parallel to the graph of $5x + 9y = 14$ must also have a slope of $-\dfrac{5}{9}$. Only Choice (B) satisfies this condition.

2. **(A)** Subtract 5 from each side of $5 - 7x \geq -9$, so that $5 - 7x - 5 \geq -9 - 5$, which simplifies to $-7x \geq -14$. Next, divide each side by -7. Since we are dividing by a negative number, we must switch the order of the inequality. Then $\dfrac{-7x}{-7} \leq \dfrac{-14}{-7}$, which becomes $x \leq 2$. This is equivalent to $X = \{x \mid x \leq 2\}$.

3. **(A)** Any number raised to a negative exponent is equivalent to the reciprocal of this number raised to the corresponding positive exponent. Then

   $4^{-1} = \left(\dfrac{1}{4}\right)^1 = \dfrac{1}{4}$ and $(-3)^{-2} = \left(-\dfrac{1}{3}\right)^2 = \dfrac{1}{9}$. Thus, $\left(\dfrac{-3}{4^{-1}}\right)(-3)^{-2} = \left(\dfrac{-3}{\frac{1}{4}}\right)\left(\dfrac{1}{9}\right) = (-3)$

   $(4)\left(\dfrac{1}{9}\right) = -\dfrac{12}{9} = -\dfrac{4}{3}$.

4. **(C)** Using one of the rules of logarithms, $\log_{10} PQ = \log_{10} P + \log_{10} Q$. Using another rule of logarithms, $\log_{10} Q^2 = 2\log_{10} Q$, which is equivalent to $\dfrac{\log_{10} Q^2}{2} = \log_{10} Q$. Thus, $\log_{10} PQ = \log_{10} P + \dfrac{\log_{10} Q^2}{2} = 0.5 + \dfrac{1.2}{2} = 0.5 + 0.6 = 1.1$.

5. **(D)** Subtract 6 from each part of the double inequality to get $-6 - 6 < \dfrac{2}{3}x + 6 - 6 < 2 - 6$, which simplifies to $-12 < \dfrac{2}{3}x < -4$. Now divide each expression by $\dfrac{2}{3}$, which is equivalent to multiplying by $\dfrac{3}{2}$. Thus, $(-12)\left(\dfrac{3}{2}\right) < \left(\dfrac{2}{3}x\right)\left(\dfrac{3}{2}\right) < (-4)\left(\dfrac{3}{2}\right)$, which leads to $-18 < x < -6$.

6. **(D)** Using the Distributive Law, $-5i^7 + (2i)(i^2 - 2) = -5i^7 + 2i^3 - 4i$. Recall that the powers of i are cyclical in groups of 4, which means that $i = i^5 = i^9 = ..., -1 = i^2 = i^6 = i^{10} = ..., -i = i^3 = i^7 = i^{11} = ...,$ and $1 = i^4 = i^8 = i^{12} = ...$ Thus, $-5i^7 + 2i^3 - 4i = (-5)(-i) + (2)(-i) - 4i = 5i - 2i - 4i = -i$.

7. **(A)** Based on the Factor Theorem, if $x + 2$ is a factor of $3x^3 - x + k$, then -2 must be a solution for x in the equation $3x^3 - x + k = 0$. This means that $(3)(-2)^3 - (-2) + k = 0$. This equation simplifies to $(3)(-8) + 2 + k = 0$. Thus, $k = 24 - 2 = 22$.

8. **(B)** The length of the first piece is x and the length of the second piece is $4x$. The length of the third piece is found by adding the first two pieces and subtracting this sum from the total of 28 inches. The sum of the first two pieces is $5x$, so the length of the third piece must be $28 - 5x$.

9. **(B)** We can eliminate the variable x from the given system of equations as follows: Multiply the first equation by 3 to get $12x + 6y = 15$. Next, multiply the second equation by 4 to get $12x + 20y = 36$. Now, by the subtraction property of equalities, $(12x + 6y) - (12x + 20y) = 15 - 36$. This equation simplifies to $-14y = -21$. Thus, $y = \dfrac{-21}{-14} = \dfrac{3}{2}$.

10. **(C)** By definition, the expression $\log_b x = y$ is equivalent to $b^y = x$. Substituting 2 for b, 32 for x, and 5 for y, we conclude that $2^5 = 32$ is equivalent to $\log_2 32 = 5$.

CHAPTER 5

Functions and Their Graphs

FUNCTIONS AND THEIR GRAPHS

ELEMENTARY FUNCTIONS

A **function** is any process that assigns a single value of y to each number of x. Because the value of x determines the value of y, y is called the **dependent variable** and x is called the **independent variable**. The set of all the values of x for which the function is defined is called the **domain** of the function. The set of corresponding values of y is called the **range** of the function.

PROBLEM

Is $y^2 = x$ a function?

SOLUTION

Graph the equation. Note that x can have two values of y. Therefore, $y^2 = x$ is not a function.

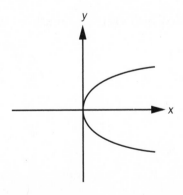

PROBLEM

Find the domain and range for $y = 5 - x^2$.

SOLUTION

First determine if there are any values that would make the function undefined (i.e., division by 0). There are none. Thus, the domain is the set of real numbers. The range can be found by substituting some corresponding values for x in the equation.

x	2	1	0	−1	−2
y	1	4	5	4	1

The range is the set of real numbers less than or equal to 5.

PROBLEM

Evaluate $f(1)$ for $y = f(x) = 5x + 2$.

SOLUTION

$f(x) = 5x + 2$

$f(1) = 5(1) + 2$

$\quad = 5 + 2$

$\quad = 7$

OPERATIONS ON FUNCTIONS

Functions can be added, subtracted, multiplied, or divided to form new functions.

a. $(f + g)(x) = f(x) + g(x)$

b. $(f - g)(x) = f(x) - g(x)$

c. $(f \times g)(x) = f(x)\,g(x)$

d. $\left(\dfrac{f}{g}\right)(x) = \dfrac{f(x)}{g(x)}, g(x) \neq 0$

PROBLEM

Let $f(x) = 2x^2 - 1$ and $g(x) = 5x + 3$. Determine the following functions:

1. $f + g$

2. $f - g$

3. $f \times g$

4. $\dfrac{f}{g}$

SOLUTION

1. $(f + g)(x) = f(x) + g(x) = 2x^2 - 1 + 5x + 3$

$$= 2x^2 + 5x + 2$$

2. $(f - g)(x) = f(x) - g(x) = 2x^2 - 1 - (5x + 3)$

$$= 2x^2 - 1 - 5x - 3$$

$$= 2x^2 - 5x - 4$$

3. $(f \times g)(x) = f(x)\,g(x) = (2x^2 - 1)(5x + 3)$

$$= 10x^3 + 6x^2 - 5x - 3$$

4. $\left(\dfrac{f}{g}\right)(x) = \dfrac{f(x)}{g(x)} = \dfrac{(2x^2 - 1)}{(5x + 3)}$

Note: The domain of (4) is for all real numbers except $-\dfrac{3}{5}$ because the fraction is indeterminate if the denominator $= 0$.

COMPOSITE FUNCTION

The **composite function** $f \circ g$ is defined $(f \circ g)(x) = f(g(x))$.

PROBLEM

Given $f(x) = 3x$ and $g(x) = 4x + 2$.

Find $(f \circ g)(x)$ and $(g \circ f)(x)$.

SOLUTION

$$(f \circ g)(x) = f(g(x)) = 3(4x + 2) = 12x + 6$$
$$(g \circ f)(x) = g(f(x)) = 4(3x) + 2 = 12x + 2$$

Note that $(f \circ g)(x) \neq (g \circ f)(x)$.

PROBLEM

Find $(f \circ g)(2)$ if $f(x) = x^2 - 3$ and $g(x) = 3x + 1$.

SOLUTION

$$(f \circ g)(2) = f(g(2))$$
$$g(x) = 3x + 1$$

Substitute the value of $x = 2$ in $g(x)$:

$$g(2) = 3(2) + 1 = 7$$
$$f(x) = x^2 - 3$$

Substitute the value of $g(2)$ in $f(x)$:

$$f(7) = (7)^2 - 3 = 49 - 3 = 46$$

INVERSE

The **inverse** of a function, f^{-1}, is obtained from f by interchanging the x and $y \, (= f(x))$ and then solving for y.

Two functions f and g are inverses of one another if $g \circ f = x$ and $f \circ g = x$. To find g when f is given, interchange x and g in the equation $y = f(x)$ and solve for $y = g(x)$.

PROBLEM

Find the inverse of the functions.

1. $f(x) = 3x + 2$

2. $f(x) = x^2 - 3$

SOLUTION

1. $f(x) = y = 3x + 2$

To find $f^{-1}(x)$, interchange x and y.

$$x = 3y + 2$$
$$3y = x - 2$$

Solve for y:

$$y = \frac{x - 2}{3}$$

2. $f(x) = y = x^2 - 3$

To find $f^{-1}(x)$, interchange x and y.

$$x = y^2 - 3$$
$$y^2 = x + 3$$

Solve for y:

$$y = \sqrt{x+3}$$

TRANSLATIONS, REFLECTIONS, AND SYMMETRY OF FUNCTIONS

A **translation** of a function will move each point of the function a specific number of units left or right, then up or down.

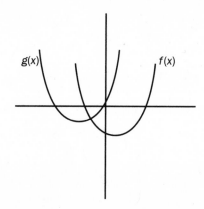

PROBLEM

Given the function $f(x) = x^2 - 5$, what is the resulting function $g(x)$ in which each point of $f(x)$ is moved 2 units to the left and 1 unit up?

SOLUTION

Since each point is moved 2 units to the left, for $g(x)$ we replace each x in $f(x)$ by $x + 2$. Also, because each point is moved 1 unit up, we add 1 to the expression for $f(x)$ to get $g(x)$. Thus, $g(x) = (x + 2)^2 - 5 + 1 = (x + 2)^2 - 4$.

To verify that this result is correct, let's choose a point on $f(x)$, such as $(3, 4)$. Following the rules for obtaining $g(x)$, the corresponding point on $g(x)$ is $(1, 5)$. Note that, by substitution, $(1, 5)$ does lie on the graph of $g(x) = (x + 2)^2 - 4$.

> A 90° **rotation** of a function moves each point P to a new point P' so that $OP = OP'$ and $\overline{OP}$ is perpendicular to $\overline{OP'}$.

The letter O represents the origin, which is located at $(0, 0)$. If the rotation is *counterclockwise*, each point (x, y) becomes $(-y, x)$. If the rotation is *clockwise*, each point (x, y) becomes $(y, -x)$.

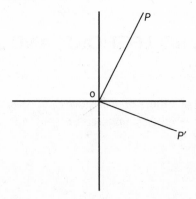

PROBLEM

Suppose a function contains the point $(4, 3)$. What are the new coordinates of this point if the function is rotated 90° counterclockwise?

SOLUTION

$(4, 3)$ will become $(-3, 4)$.

PROBLEM

Suppose a function contains the point (4, 3). What are the new coordinates of (4, 3) if the function is rotated 90° clockwise?

SOLUTION

(4, 3) will become (3, −4).

> A **reflection** of a function is simply the mirror image of the function.

A reflection about the x-axis changes point (x, y) into point $(x, -y)$. A reflection about the y-axis changes point (x, y) into point $(-x, y)$. A reflection about the line $y = x$ will move each point P to a new point P' so that the line $y = x$ is the perpendicular bisector of $\overline{PP'}$. Each point (x, y) becomes (y, x) after the reflection.

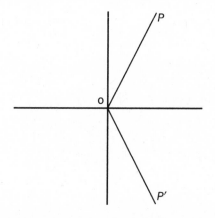

PROBLEM

A function contains the point (−6, 3). If this function is reflected about the line $y = x$, what will be the new coordinates for (−6, 3)?

SOLUTION

(−6, 3) will become (3, −6).

Drill Questions

1. If $f(x) = 2x + 3$ and $g(x) = x^3 - 5$, what is the value of $f(g(-3))$?

   (A) -61
   (B) -32
   (C) 25
   (D) 47

2. If $h(x) = \dfrac{5}{x - 2}$, which of the following is equivalent to the inverse of $h(x)$?

   (A) $\dfrac{x - 2}{5}$

   (B) $\dfrac{5}{x + 2}$

   (C) $\dfrac{2x - 5}{5}$

   (D) $\dfrac{2x + 5}{x}$

3. What is the domain of $f(x) = \sqrt{6 - 3x}$?

   (A) $x \geq 2$
   (B) $x \leq 2$
   (C) $x \geq -2$
   (D) $x \leq -2$

4. If $\{(4, -1), (5, -3), (6, -9), (\_\_, -16)\}$ represents a function, which one of the following *cannot* be filled in the blank space?

   (A) 5
   (B) 3
   (C) -4
   (D) -9

5. Each point of the function $f(x) = x^2 + 10$ is moved 3 units to the left and 2 units down to create a new function $g(x)$. What would be the y-coordinate of a point on the graph of $g(x)$ whose x-coordinate is -1 on the graph of $f(x)$?

   (A) 6
   (B) 7
   (C) 8
   (D) 9

6. If $f(x)$ is a linear function such that $f(1) = 5$ and $f(4) = -7$, then what is the value of $f(-1)$?

   (A) 11
   (B) 12
   (C) 13
   (D) 14

7. The function $g(x)$ is known to have a range of all real numbers except zero? Which one of the following expressions could represent $g(x)$?

   (A) $4x^2 - 4$

   (B) $\dfrac{4}{x}$

   (C) $\dfrac{4}{x^2}$

   (D) $-\sqrt{x^2 - 4}$

8. The point $(5, -8)$ is reflected across the line $y = x$. What is the new location?

   (A) $(-5, 8)$
   (B) $(-5, -8)$
   (C) $(-8, 5)$
   (D) $(-8, -5)$

9. The point $(-2, -3)$ is rotated $90°$ clockwise about the origin. What is its new location?

   (A) $(-3, 2)$
   (B) $(-3, -2)$
   (C) $(3, -2)$
   (D) $(3, 2)$

10. For which one of the following functions is $f(1) = f(-1) = 2$?

(A) $f(x) = x^2 - 3x + 4$
(B) $f(x) = 4x^2 - 2$
(C) $f(x) = x^2 - 2x - 1$
(D) $f(x) = 3x^2 - x$

Answers to Drill Questions

1. **(A)** $g(-3) = (-3)^3 - 5 = -27 - 5 = -32$. Thus, $f(g(-3)) = f(-32) = (2)(-32) + 3 = -64 + 3 = -61$.

2. **(D)** The inverse of a function is found by reversing the roles of the two variables, then solving for the new dependent variable. Let $y = h(x)$ so that the initial function can be written as $y = \dfrac{5}{x-2}$. Interchanging the x and y, we get $x = \dfrac{5}{y-2}$. Multiply this equation by $y - 2$ to get $(x)(y-2) = 5$. This equation simplifies to $xy - 2x = 5$. Next, add $2x$ to each side to get $xy - 2x + 2x = 5 + 2x$, which becomes $xy = 5 + 2x$. Finally, divide both sides by x to get $y = \dfrac{5+2x}{x}$. The right side of this equation, which is equivalent to $\dfrac{2x+5}{x}$, represents the expression for the inverse of $h(x)$.

3. **(B)** The domain of $f(x) = \sqrt{6 - 3x}$ is defined as the allowable values of x. The square root of any function must be at least zero in order to represent a real value. The domain is found by solving $6 - 3x \geq 0$. Subtracting 6 from each side leads to $-3x \geq -6$. Finally, divide both sides by -3 and reverse the order of the inequality. Thus, $\dfrac{-3x}{-3} \leq \dfrac{-6}{-3}$, which becomes $x \leq 2$.

4. **(A)** The definition of a function, as it relates to a set of ordered pairs, is that any specific first number must be associated with a single second number. This implies that a set of ordered pairs is not a function if two ordered pairs contain the same first number but a different second number. The given set contains the elements $(4, -1)$, $(5, -3)$, $(6, -9)$, and $(\_, -16)$. In order for this set to qualify as a function, we cannot repeat 4, 5, or 6 as a first number of the last ordered pair.

5. **(D)** Substitute -1 for x in the function $f(x)$ to get $(-1)^2 + 10 = 1 + 10 = 11$. The corresponding point on the graph of $f(x)$ is $(-1, 11)$. In order to find the

1

corresponding point for $g(x)$, this point will be moved 3 units to the left and 2 units down. Thus, the point $(-1, 11)$ will become the point $(-4, 9)$ on the graph of $g(x)$. So, the y-coordinate becomes 9.

6. **(C)** The points on the graph of this linear function are $(1, 5)$ and $(4, -7)$. The function can be written as $y = mx + b$, where m is the slope and b is the y-intercept. By substitution of the two given points, we get $5 = (m)(1) + b$ and $-7 = (m)(4) + b$. Now subtracting the second of these equations from the first equation leads to $5 - (-7) = (m)(1) + b - (m)(4) - b$, which becomes $12 = -3m$. So $m = -4$. Returning to the equation $5 = (m)(1) + b$ we can substitute -4 for m so that $5 = (-4)(1) + b$. Then $b = 9$. The equation of this linear function is $y = -4x + 9$. Let's replace y with $f(x)$. Finally, $f(-1) = (-4)(-1) + 9 = 4 + 9 = 13$.

7. **(B)** The range is represented by the $g(x)$ values. For $g(x) = \dfrac{4}{x}$, $g(x)$ can assume any value (including negative numbers) except zero. Also, note that $x \neq 0$. A graph of $g(x) = \dfrac{4}{x}$ would also confirm that the range is all numbers except zero.

8. **(C)** If the point (x, y) is reflected across the line $y = x$, its new location is given by (y, x). Let $x = 5$ and $y = -8$. Then for the point $(5, -8)$, its new location after being reflected across the line $y = x$ is found by simply interchanging the coordinates. The correct answer is $(-8, 5)$.

9. **(A)** If the point (x, y) is rotated 90° clockwise about the origin, its new location is given by $(y, -x)$. Let $x = -2$ and $y = -3$. Then for the point $(-2, -3)$, its new location after being rotated 90° clockwise about the origin is found by interchanging the coordinates, then switching the sign of the new second coordinate. The correct answer is $(-3, 2)$.

10. **(B)** The quickest way to solve this problem is by substitution into each answer choice. For choice (A), $f(1) = 1^2 - (3)(1) + 4 = 2$, but $f(-1) = (-1)^2 - (3)(-1) + 4 = 8$. So, choice (A) is incorrect. Choice (B) is correct because $f(1) = 4(1)^2 - 2 = 2$ and $f(-1) = 4(-1)^2 - 2 = 2$. Note that for choice (C), $f(1) = -2$, but $f(-1) = 2$. For choice (D), $f(1) = 2$, but $f(-1) = 4$.

CHAPTER 6

Geometry Topics

GEOMETRY TOPICS

Plane geometry refers to two-dimensional shapes (that is, shapes that can be drawn on a sheet of paper), such as triangles, parallelograms, trapezoids, and circles. Three-dimensional objects (that is, shapes with depth) are the subjects of solid geometry.

TRIANGLES

A closed three-sided geometric figure is called a **triangle**. The points of the intersection of the sides of a triangle are called the **vertices** of the triangle.

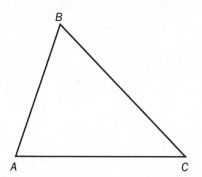

A **side** of a triangle is a line segment whose endpoints are the vertices of two angles of the triangle. The perimeter of a triangle is the sum of the measures of the sides of the triangle.

An **interior angle** of a triangle is an angle formed by two sides and includes the third side within its collection of points. The sum of the measures of the interior angles of a triangle is 180°.

A **scalene triangle** has no equal sides.

An **isosceles triangle** has at least two equal sides. The third side is called the **base** of the triangle, and the base angles (the angles opposite the equal sides) are equal.

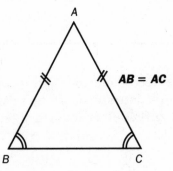

$AB = AC$

An **equilateral triangle** has all three sides equal. $\overline{AB} = \overline{AC} = \overline{BC}$. An equilateral triangle is also **equiangular**, with each angle equaling 60°.

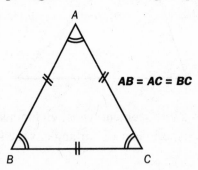

$AB = AC = BC$

An **acute triangle** has three acute angles (less than 90°).

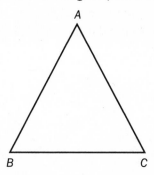

An **obtuse triangle** has one obtuse angle (greater than 90°).

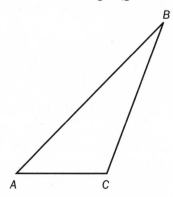

A **right triangle** has a right angle. The side opposite the right angle in a right triangle is called the **hypotenuse** of the right triangle. The other two sides are called the **legs** (or arms) of the right triangle. By the **Pythagorean Theorem**, the lengths of the three sides of a right triangle are related by the formula

$$c^2 = a^2 + b^2$$

where c is the hypotenuse and a and b are the other two sides (the legs). The Pythagorean Theorem is discussed in more detail in the next section.

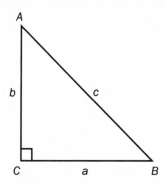

An **altitude**, or **height**, of a triangle is a line segment from a vertex of the triangle perpendicular to the opposite side. For an obtuse triangle, the altitude sometimes is drawn as a perpendicular line to an extension of the opposite side.

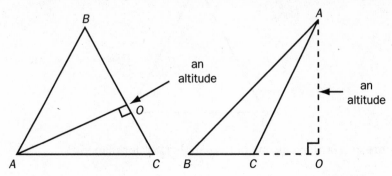

The **area** of a triangle is given by

$$A = \frac{1}{2}bh$$

where h is the altitude and b is the base to which the altitude is drawn.

A line segment connecting a vertex of a triangle and the midpoint of the opposite side is called a **median** of the triangle.

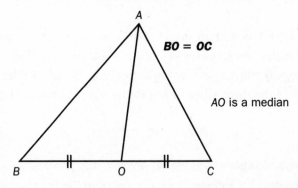

A line that bisects and is perpendicular to a side of a triangle is called a **perpendicular bisector** of that side.

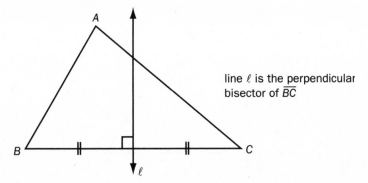

line ℓ is the perpendicular bisector of $\overline{BC}$

An **angle bisector** of a triangle is a line that bisects an angle and extends to the opposite side of the triangle.

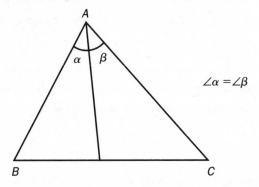

$\angle\alpha = \angle\beta$

The line segment that joins the midpoints of two sides of a triangle is called a **midline** of the triangle.

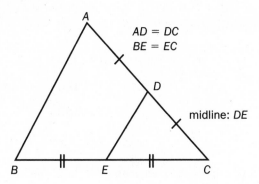

AD = DC
BE = EC

midline: DE

An **exterior angle** of a triangle is an angle formed outside a triangle by one side of the triangle and the extension of an adjacent side.

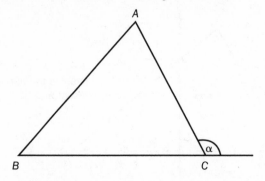

PROBLEM

The measure of the vertex angle of an isosceles triangle exceeds the measure of each base angle by 30°. Find the value of each angle of the triangle.

SOLUTION

In an isosceles triangle, the angles opposite the congruent sides (the base angles) are, themselves, congruent and of equal value.

Therefore,

1. Let x = the measure of each base angle

2. Then $x + 30$ = the measure of the vertex angle

We can solve for x algebraically by keeping in mind that the sum of all the measures of the angles of a triangle is 180°.

$$x + x + (x + 30) = 180$$
$$3x + 30 = 180$$
$$3x = 150$$
$$x = 50$$

Therefore, the base angles each measure 50°, and the vertex angle measures 80°.

THE PYTHAGOREAN THEOREM

The **Pythagorean Theorem** pertains to a right triangle, which, as we saw, is a triangle that has one 90° angle. The Pythagorean Theorem tells you that the square of the hypotenuse of a right triangle is equal to the sum of the squares of the other two sides, or

$$c^2 = a^2 + b^2$$

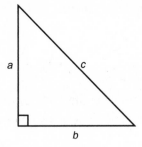

The Pythagorean Theorem is useful because if you know the length of any two sides of a right triangle, you can figure out the length of the third side.

PROBLEM

In a right triangle, one leg is 3 inches and the other leg is 4 inches. What is the length of the hypotenuse?

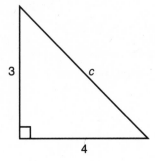

SOLUTION

$$c^2 = a^2 + b^2$$
$$c^2 = 3^2 + 4^2$$
$$c^2 = 9 + 16$$
$$c^2 = 25$$
$$c = 5$$

PROBLEM

If one leg of a right triangle is 6 inches, and the hypotenuse is 10, what is the length of the other leg?

SOLUTION

First, write down the equation for the Pythagorean Theorem. Next, plug in the information you are given. The hypotenuse c is equal to 10 and one of the legs, $b,$ is equal to 6. Solve for a.

$$c^2 = a^2 + b^2$$
$$a^2 = c^2 - b^2$$
$$a^2 = 10^2 - 6^2$$
$$a^2 = 100 - 36$$
$$a^2 = 64$$
$$a = 8 \text{ inches}$$

PROBLEM

What is the value of b in the right triangle shown below?

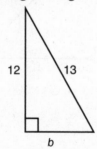

12 13

b

SOLUTION

To answer this question, you need to use the Pythagorean Theorem. The problem is asking for the value of the missing leg.

$$c^2 = a^2 + b^2$$
$$b^2 = c^2 - a^2$$
$$b^2 = 13^2 - 12^2$$
$$b^2 = 169 - 144$$
$$b^2 = 25$$
$$b = 5$$

QUADRILATERALS

> A **polygon** is any closed figure with straight line segments as sides. A **quadrilateral** is any polygon with four sides. The points where the sides meet are called **vertices** (singular: **vertex**).

PARALLELOGRAMS

A **parallelogram** is a quadrilateral whose opposite sides are parallel.

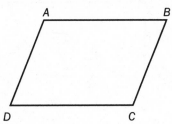

Two angles that have their vertices at the endpoints of the same side of a parallelogram are called **consecutive angles**. So $\angle A$ is consecutive to $\angle B$; $\angle B$ is consecutive to $\angle C$; $\angle C$ is consecutive to $\angle D$; and $\angle D$ is consecutive to $\angle A$.

The perpendicular segment connecting any point of a line containing one side of a parallelogram to the line containing the opposite side of the parallelogram is called the **altitude** of the parallelogram.

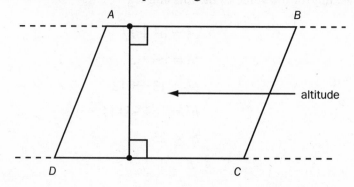

A **diagonal** of a polygon is a line segment joining any two nonconsecutive vertices. The area of a parallelogram is given by the formula $A = bh$, where b is the base and h is the height drawn perpendicular to that base. Note that the height is the same as the altitude of the parallelogram.

Example:

The area of the parallelogram below is:

$$A = bh$$

$$A = (10)(3)$$

$$A = 30$$

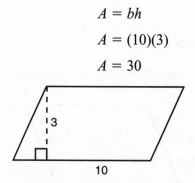

RECTANGLES

A **rectangle** is a parallelogram with right angles.

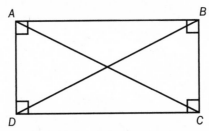

- The diagonals of a rectangle are equal, $\overline{AC} = \overline{BD}$.

- If the diagonals of a parallelogram are equal, the parallelogram is a rectangle.

- If a quadrilateral has four right angles, then it is a rectangle.

- The area of a rectangle is given by the formula $A = lw$, where l is the length and w is the width.

Example:

The area of the rectangle below is:

$$A = lw$$

$$A = (4)(9)$$

$$A = 36$$

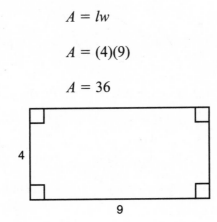

RHOMBI

A **rhombus** (plural: **rhombi**) is a parallelogram that has two adjacent sides that are equal.

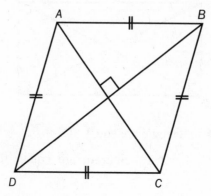

- All sides of a rhombus are equal.

- The diagonals of a rhombus are perpendicular bisectors of each other.

- The area of a rhombus can be found by the formula $A = \frac{1}{2}(d_1 \times d_2)$, where d_1 and d_2 are the diagonals.

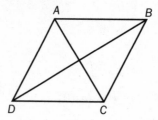

$ABCD$ is a rhombus. $AC = 4$ and $BD = 7$. The area of the rhombus is $\left(\frac{1}{2}\right)(AC)(BD) = \left(\frac{1}{2}\right)(4)(7) = 14$.

- The diagonals of a rhombus bisect the angles of the rhombus.

- If the diagonals of a parallelogram are perpendicular, the parallelogram is a rhombus.

- If a quadrilateral has four equal sides, then it is a rhombus.

- A parallelogram is a rhombus if either diagonal of the parallelogram bisects the angles of the vertices it joins.

SQUARES

A **square** is a rhombus with a right angle.

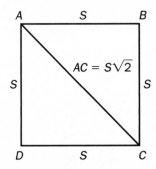

- A square is an equilateral quadrilateral.

- A square has all the properties of rhombi and rectangles.

- In a square, the measure of either diagonal can be calculated by multiplying the length of any side by the square root of 2.

- The area of a square is given by the formula $A = s^2$, where s is the side of the square.

- Since all sides of a square are equal, it does not matter which side is used.

Example:

The area of the square shown below is:

$$A = s^2$$

$$A = 6^2$$

$$A = 36$$

6

The area of a square can also be found by taking $\frac{1}{2}$ the product of the length of the diagonal squared. This comes from a combination of the facts that the area of a rhombus is $\left(\frac{1}{2}\right) d_1 d_2$ and that $d_1 = d_2$ for a square.

Example:

The area of the square shown below is:

$$A = \frac{1}{2}d^2$$

$$A = \frac{1}{2}(8)^2$$

$$A = 32$$

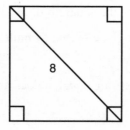

TRAPEZOIDS

A **trapezoid** is a quadrilateral with two and only two parallel sides. The parallel sides of a trapezoid are called the **bases**. The **median** of a trapezoid is the line joining the midpoints of the nonparallel sides.

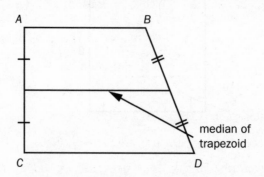

median of trapezoid

The perpendicular segment connecting any point in the line containing one base of the trapezoid to the line containing the other base is the **altitude** of the trapezoid.

A pair of angles including only one of the parallel sides is called a pair of **base angles**.

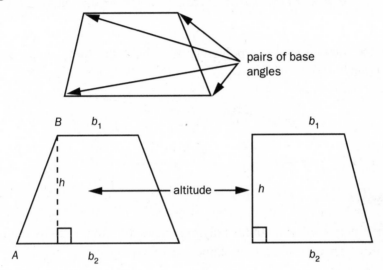

pairs of base angles

- The median of a trapezoid is parallel to the bases and equal to one-half their sum.

- The area of a trapezoid equals one-half the altitude times the sum of the bases, or $\frac{1}{2}h(b_1 + b_2)$.

- An **isosceles trapezoid** is a trapezoid whose non-parallel sides are equal. A pair of angles including only one of the parallel sides is called a pair of base angles.

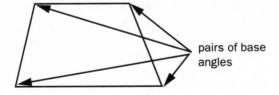

pairs of base angles

- The base angles of an isosceles trapezoid are equal.

- The diagonals of an isosceles trapezoid are equal.

- The opposite angles of an isosceles trapezoid are supplementary.

SIMILAR POLYGONS

> Two polygons are **similar** if there is a one-to-one corre-spondence between their vertices such that all pairs of corresponding angles are congruent and the ratios of the measures of all pairs of corresponding sides are equal.

Note that although similar polygons must have the same shape, they may have different sizes.

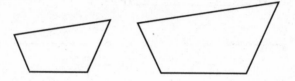

Theorem 1

The perimeters of two similar polygons have the same ratio as the measure of any pair of corresponding line segments of the polygons.

Theorem 2

The ratio of the lengths of two corresponding diagonals of two similar polygons is equal to the ratio of the lengths of any two corresponding sides of the polygons.

Theorem 3

The areas of two similar polygons have the same ratio as the square of the measures of any pair of corresponding sides of the polygons.

Theorem 4

Two polygons composed of the same number of triangles similar to each, and similarly placed, are similar. Thus, $ABCD$ is similar to $A'B'C'D'$. Note that when naming similar polygons, the corresponding letters must match: A to A', B to B', C to C', and D to D'.

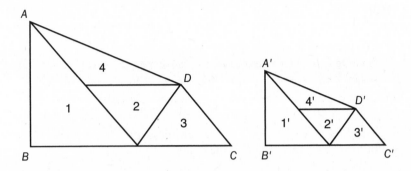

PROBLEM

The lengths of two corresponding sides of two similar polygons are 4 and 7. If the perimeter of the smaller polygon is 20, find the perimeter of the larger polygon.

SOLUTION

We know, by theorem, that the perimeters of two similar polygons have the same ratio as the measures of any pair of corresponding sides.

If we let s and p represent the side and perimeter of the smaller polygon and s' and p' represent the corresponding side and perimeter of the larger one, we can then write the proportion

$$s : s' = p : p'; \text{ or } \frac{s}{s'} = \frac{p}{p'}$$

By substituting the given values, we can solve for p'.

$$\frac{4}{7} = \frac{20}{p'}$$
$$4p' = 140$$
$$p' = 35$$

Therefore, the perimeter of the larger polygon is 35.

CIRCLES

A **circle** is a set of points in the same plane equidistant from a fixed point, called its **center**. Circles are often named by their center point, such as circle **O** below.

A **radius** of a circle is a line segment drawn from the center of the circle to any point on the circle.

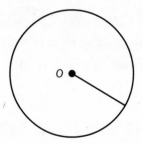

The **circumference** of a circle is the length of its outer edge, given by

$$C = \pi d = 2\pi r$$

where r is the radius, d is the diameter, and π (pi) is a mathematical constant approximately equal to 3.14.

The **area** of a circle is given by

$$A = \pi r^2$$

A full circle is 360°. The measure of a semicircle (half a circle) is 180°.

A line that intersects a circle in two points is called a **secant**.

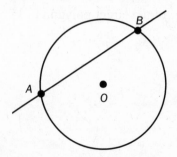

A line segment joining two points on a circle is called a **chord** of the circle.

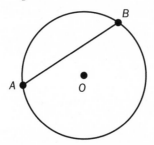

A chord that passes through the center of the circle is called a **diameter** of the circle. The length of the diameter is twice the length of the radius, $d = 2r$.

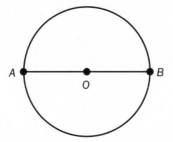

The line passing through the centers of two (or more) circles is called the **line of centers**.

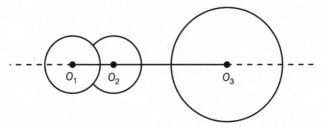

An angle whose vertex is on the circle and whose sides are chords of the circle is called an **inscribed angle** ($\angle BAC$ in the diagrams).

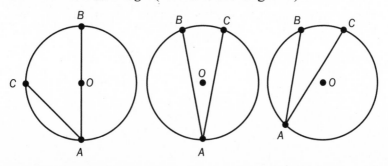

An angle whose vertex is at the center of a circle and whose sides are radii is called a **central angle**. The portion of a circle cut off by a central angle is called an **arc** of the circle.

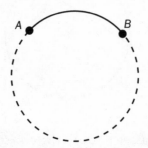

The measure of a minor arc is the measure of the central angle that intercepts that arc. The measure of a semicircle (half a circle) is 180°.

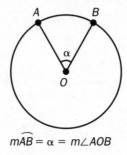

$$m\overset{\frown}{AB} = \alpha = m\angle AOB$$

The length of an arc intercepted by a central angle has the same ratio to the circle's circumference as the measure of the arc is to 360°, the full circle. Therefore, arc length is given by $\dfrac{n}{360} \times 2\pi r$, where n = measure of the central angle.

A sector is the portion of a circle between two radii (sector *AOB* here). Its area is given by $A = \dfrac{n}{360}(\pi r^2)$, where n is the central angle formed by the radii.

The distance from an outside point P to a given circle is the distance from that point to the point where the circle intersects with a line segment with endpoints at the center of the circle and point P. The distance of point P to the diagrammed circle with center O is the line segment $\overline{PB}$, part of line segment $\overline{PO}$.

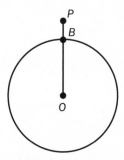

A line that has one and only one point of intersection with a circle is called a **tangent** to that circle, and their common point is called a **point of tangency**.

In the diagram, Q and P are each points of tangency. A tangent is always perpendicular to the radius drawn to the point of tangency.

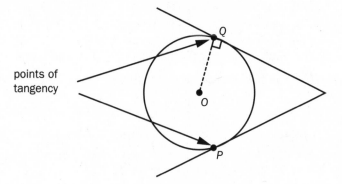

Congruent circles are circles whose radii are congruent.

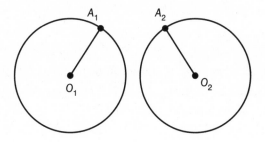

If $O_1A_1 \cong O_2A_2$, then $O_1 \cong O_2$.

Circles that have the same center and unequal radii are called **concentric circles**.

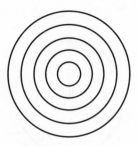

A **circumscribed circle** is a circle passing through all the vertices of a polygon. The polygon is said to be **inscribed** in the circle.

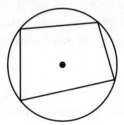

PROBLEM

A and *B* are points on a circle *Q* such that $\triangle AQB$ is equilateral. If the length of side $\overline{AB} = 12$, find the length of arc *AB*.

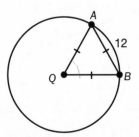

SOLUTION

To find the length of arc *AB*, we must find the measure of the central angle $\angle AQB$ and the measure of radius $\overline{QA}$. $\angle AQB$ is an interior angle of the equilateral triangle $\triangle AQB$. Therefore, $m \angle AQB = 60°$.

Similarly, in the equilateral $\triangle AQB$, $\overline{AQ} = \overline{AB} = \overline{QB} = 12 = r$.

Given the radius, *r*, and the central angle, *n*, the arc length is given by

$$\frac{n}{360} \times 2\pi r = \frac{60}{360} \times 2\pi \times 12 = \frac{1}{6} \times 2\pi \times 12 = 4\pi.$$

Therefore, the length of arc $AB = 4\pi$.

FORMULAS FOR AREA AND PERIMETER

| **Figures** | **Areas** |

Area (A) of a:

square	$A = s^2$; where s = side
rectangle	$A = lw$; where l = length, w = width
parallelogram	$A = bh$; where b = base, h = height
triangle	$A = \dfrac{1}{2}bh$; where b = base, h = height
circle	$A = \pi r^2$; where $\pi = 3.14$, r = radius
sector	$A = \left(\dfrac{n}{360}\right)(\pi r^2)$; where n = central angle, r = radius, $\pi = 3.14$
trapezoid	$A = \left(\dfrac{1}{2}\right)(h)(b_1 + b_2)$; where h = height, b_1 and b_2 = bases

| **Figures** | **Perimeters** |

Perimeter (P) of a:

square	$P = 4s$; where s = side
rectangle	$P = 2l + 2w$; where l = length, w = width
triangle	$P = a + b + c$; where a, b, and c are the sides
Circumference (C) of a circle	$C = \pi d$; where $\pi = 3.14$, d = diameter

PROBLEM

Points P and R lie on circle Q, $m \angle PQR = 120°$, and $PQ = 18$. What is the area of sector PQR?

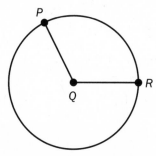

SOLUTION

$$\frac{120°}{360°} = \frac{\text{Area of sector } PQR}{\text{Area of circle } Q}$$

Letting X = area of sector PQR, and replacing area of circle Q with $\pi(18^2)$ = 324π, we get

$$\frac{120°}{360°} = \frac{X}{324\pi}$$

$$\text{Then } X = \frac{(120°)(324\pi)}{360°} = 108\pi$$

Drill Questions

1. What is the area of the following right triangle?

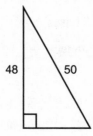

(A) 1200
(B) 672
(C) 336
(D) 112

2. Parallelogram $RSTU$ is similar to parallelogram $WXYZ$. If $\angle RST = 60°$, $\angle XYZ =$

(A) 60°
(B) 90°
(C) 120°
(D) Not enough information is given

3. The center of circle O is at the origin, as shown. Point (2, 2) is on the circle. What is the circumference of circle O?

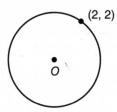
(2, 2)
O

(A) 4π
(B) $4\sqrt{2}\pi$
(C) 2π
(D) $2\sqrt{2}\pi$

4. If the short side of a rectangle measures 3 inches, and its long side is twice as long, what is the length of a diagonal of a square with the same area as this rectangle?

(A) $3\sqrt{2}$
(B) 6
(C) 18
(D) 36

5. Which of the following *cannot* be the lengths of the sides of a triangle?

(A) 2, 5, 6
(B) 3, 4, 5
(C) 4, 5, 6
(D) 4, 5, 10

6. Find the length of the missing side in this right triangle.

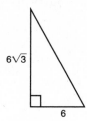
$6\sqrt{3}$
6

(A) 6
(B) $\sqrt{66}$
(C) 12
(D) $\sqrt{306}$

7. Given the intersecting lines and angle measurement in the figure, $x =$

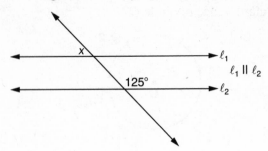

(A) 55°
(B) 60°
(C) 65°
(D) 70°

8. This figure shows sector *AOB* equal to a quarter of a circle, circumscribed about a square. What is the area of the shaded region?

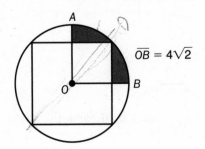

$$\overline{OB} = 4\sqrt{2}$$

(A) $16(2\pi - 1)$
(B) $8\sqrt{2\pi} - 16$
(C) 8π
(D) $8(\pi - 2)$

9. An isosceles right triangle has angles of

(A) 30°, 60°, 90°
(B) 45°, 45°, 90°
(C) 0°, 90°, 90°
(D) 60°, 60°, 60°

10. An angle of measure of 180° is termed

    (A) straight
    (B) supplementary
    (C) obtuse
    (D) reflex

Answers to Drill Questions

1. **(C)** Let x represent the length of the horizontal base of this triangle. By the Pythagorean theorem, $x^2 + 48^2 = 50^2$. Then $x^2 + 2,304 = 2,500$, which means that $x^2 = 2,500 - 2,304 = 196$. So $x = \sqrt{196} = 14$. The area of the triangle is one-half the product of the base and the height. Thus, the required area is $\left(\frac{1}{2}\right)(14)(48) = \left(\frac{1}{2}\right)(672) = 336$.

2. **(C)** Corresponding angles of similar geometric figures must be congruent. This means that $\angle XYZ$ in parallelogram $WXYZ$ must be congruent to $\angle STU$ in parallelogram $RSTU$. We note that $\angle RST$ and $\angle STU$ are consecutive angles in $RSTU$ and the sum of any two consecutive angles in a parallelogram is 180°. Since $\angle RST = 60°$, $\angle STU = 180° - 60° = 120°$. Thus, $\angle XYZ = 120°$.

3. **(B)** The length of the radius of this circle is found by the distance between point O which is located at $(0, 0)$ and the point $(2, 2)$. The distance between $(0, 0)$ and $(2, 2)$ is $\sqrt{(0-2)^2 + (0-2)^2} = \sqrt{(-2)^2 + (-2)^2} = \sqrt{4+4} = \sqrt{8}$. Note that we can write $\sqrt{8}$ as $\left(\sqrt{4}\right)\left(\sqrt{2}\right) = 2\sqrt{2}$. Finally, the circumference equals the product of 2π and the radius, which is $(2\pi)\left(2\sqrt{2}\right) = 4\pi\sqrt{2} = 4\sqrt{2}\pi$.

4. **(B)** The long side of the rectangle measures $(2)(3) = 6$ inches, so its area is $(6)(3) = 18$ square inches. Then the area of the square is also 18 square inches. Then each side of the square is $\sqrt{18}$ inches. The diagonal of a square can represent the hypotenuse of a triangle whose two legs are consecutive sides of the square. Let d represent the length of the diagonal. By the Pythagorean theorem, $d^2 = \left(\sqrt{18}\right)^2 + \left(\sqrt{18}\right)^2 = 18 + 18 = 36$. Thus, $d = \sqrt{36} = 6$.

5. **(D)** The sum of any two sides of a triangle must exceed the length of the third side. This means that a triangle *cannot* have sides with lengths of 4, 5, and 10 because $4 + 5 < 10$.

6. **(C)** The missing side represents the hypotenuse of this triangle. Let x represent its length. By the Pythagorean theorem, $x^2 + 6^2 + \left(6\sqrt{3}\right)^2$. This equation simplifies to $x^2 = 36 + (36)(3) = 144$. Thus, $x = \left(\sqrt{144}\right) = 12$.

7. **(A)** When a transversal intersects two parallel lines, corresponding angles are congruent. By definition, x and the angle represented by 125° are corresponding angles. As such, $x = 125°$. Now note that the angle represented by x and 125° are adjacent angles that form a straight line. This means that $x + 125° = 180°$. Thus, $x = 180° - 125° = 55°$.

8. **(D)** The diagonal of the square $\overline{OD}$ is a radius of the circle, so its length is equal to $\overline{OB}$, which is $4\sqrt{2}$. Let x represent the length of each side of the square. By the Pythagorean theorem, $\left(\overline{OC}\right)^2 + \left(\overline{CD}\right)^2 = \left(\overline{OD}\right)^2$. Then $x^2 + x^2 = \left(4\sqrt{2}\right)^2 = (16)(2) = 32$. This equation can be simplified to $2x^2 = 32$, which further simplifies to $x^2 = 16$. This means that the area of the square is 16. The area of the quarter-circle formed by the points A, O, and B is one-fourth the area of the circle. This area is found by the expression $\left(\frac{1}{4}\right)(\pi)(r^2)$, where r is the length of the radius. Since we know the radius length to be $4\sqrt{2}$, the area of the quarter-circle is $\left(\frac{1}{4}\right)(\pi)\left(4\sqrt{2}\right)^2 = \left(\frac{1}{4}\right)(\pi)(16)(2) = \left(\frac{1}{4}\right)(\pi)(32) = 8\pi$. Finally, the area of the shaded region is the difference between the area of the quarter-circle and the area of the square, which is $8\pi - 16$. This is equivalent to $8(\pi - 2)$.

9. **(B)** An isosceles right triangle must consist of a 90°-angle and two acute congruent angles whose sum is 90°. The only set of numbers that satisfies these conditions are angle measures of 45°, 45°, 90°.

10. **(A)** By definition, an angle whose measure is 180° is called a straight angle.

CHAPTER 7

Probability and Statistics

PROBABILITY AND STATISTICS

The first part of this chapter discusses counting principles and how they relate to permutations and combinations, which are the building blocks of probability. The remainder of the chapter addresses statistics, introduces data measurements—such as mean, median, mode, and standard deviation—and reviews data analysis of various charts and graphs.

THE FUNDAMENTAL COUNTING PRINCIPLE

The **fundamental counting principle** deals with identifying the number of outcomes of a given experiment and encompasses the **counting rule**:

If one experiment can be performed in m ways, and a second experiment can be performed in n ways, then there are $m \times n$ distinct ways both experiments can be performed in this specified order. The counting principle can be applied to more than two experiments.

PROBLEM

A new line of children's clothing is color-coordinated. Niki's mother bought her five tops, three shorts, and two pairs of shoes for the summer. How many different outfits can Niki choose from?

SOLUTION

How many choices are there for a top?   5

How many choices are there for shorts?   3

How many choices are there for shoes?   2

Apply the counting principle: $5 \times 3 \times 2$ or 30.

Thus, Niki has 30 different outfits.

PERMUTATIONS

A **permutation** is an arrangement of specific objects in which order is of particular importance. To determine the number of possible permutations, the following formula can be used:

$$_nP_r = \frac{n!}{(n-r)!}$$

where n is the number of objects in the given set, r is the number of objects being chosen, and ! is the notation used for factorial. The **factorial** of a number is the product of that number and all the numbers less than it down to 1, or $n! = n \times (n-1) \times (n-2) \times ... \times 3 \times 2 \times 1$.

For example, $5! = 5 \times 4 \times 3 \times 2 \times 1 = 120$.

The formula for permutations is a consequence of the counting rule described above.

Example:

$_6P_2 = \dfrac{6!}{4!} = \dfrac{6 \times 5 \times 4 \times 3 \times 2 \times 1}{4 \times 3 \times 2 \times 1}$. Note that 4! cancels out part of 6! and we are left with only $6 \times 5 = 30$. So you really don't have to do that much multiplication. Permutations cancel down to $n \times (n-1) \times (n-2) \times ... (n-r+1)$, so even a permutation with large numbers, such as $_{12}P_3$ becomes $12 \times 11 \times 10 = 1320$.

PROBLEM

Ten children in Ms. Berea's fifth grade class compete in the school's science fair. First, second, and third place trophies will go to the top three projects. In how many ways can the trophies be awarded?

SOLUTION

Order is of definite importance here, so we use $_{10}P_3 = 10 \times 9 \times 8 = 720$. Thus, there are 720 ways these 10 children could finish first, second, or third.

Let's look at the same problem using the counting principle:

SOLUTION

How many children have a chance at the first place trophy?   10

How many children have a chance at the second place trophy?   9

How many children have a chance at the third place trophy?   8

Apply the counting principle: $10 \times 9 \times 8 = 720$.

COMBINATIONS

A **combination** is an arrangement of specific objects in which order is *not* of particular importance. To determine the number of possible combinations, use the following formula:

$$_nC_r = \frac{n!}{r!(n-r)!}$$

where *n* is the number of objects in the given set, *r* is the number of objects being ordered, and ! is the notation used for factorial.

Note that this formula is similar to the one for permutations, but since order is not important, we have to factor out all the duplications. In combinations, *ABC* is the same as *ACB*, *BAC*, *BCA*, *CAB*, and *CBA*. So we must divide the permutation by the duplications, which occur *r*! times.

As with permutations, we don't have to do too much calculation for combinations. After the cancellations, combinations reduce to a fraction in which the first *r* factors of *n*! are in the numerator and *r*! is in the denominator, and then further cancellations can take place. For example, $_8C_3 = \dfrac{8 \times 7 \times 6}{3 \times 2 \times 1}$, which cancels to $8 \times 7 = 56$. The final cancellations are always possible in combinations before multiplying to get the final answer, and the denominator *always* cancels out.

PROBLEM

Simon has 10 compact discs and his player holds three discs at a time. How many combinations of three discs are possible?

SOLUTION

$$_{10}C_3 = \frac{10!}{3!(10-3)!}$$

$$= \frac{10!}{3! \times 7!}$$

$$= \frac{10 \times 9 \times 8 \times 7 \times 6 \times 5 \times 4 \times 3 \times 2 \times 1}{(3 \times 2 \times 1) \times (7 \times 6 \times 5 \times 4 \times 3 \times 2 \times 1)}$$

Cancel the 3 into the 9 and the 2 into the 8 to get $10 \times 3 \times 4 = 120$.

Thus, there are 120 combinations of discs taken three at a time.

PROBABILITY

Before discussing the actual calculation of a probability, let's review the following probability facts:

1. Probabilities are values ranging from 0 to 1 inclusive: $0 \le P(E) \le 1$, where $P(E)$ means probability (P) of an event (E).

    a. Probabilities cannot be negative or greater than 1.

    b. A probability of 0 means that the event cannot or did not occur.

    c. A probability of 1 means that the event must occur or always occurred.

    d. A probability is expressed as a fraction or a decimal.

2. An experiment consists of all possible outcomes.

3. An event consists of one or more outcomes.

4. The sum of the probabilities of all possible outcomes in any given experiment is 1.

5. The notation used for the probability of an event E not happening is $P(E')$, where E' is read as "E complement."

6. Combining facts (4) and (5), we then see that $P(E) + P(E') = 1$. That is, the probability of an event happening or its complement happening is 1. This formula may also be applied in the following form: $P(E) = 1 - P(E')$, depending on the context of the problem.

Example:

An experiment consists of rolling an ordinary number cube once. Then, each of 1, 2, 3, 4, 5, 6 are outcomes and $P(1) + P(2) + P(3) + P(4) + P(5) + P(6) = 1$. An example of an event E_1 would be the set of outcomes numbered below 3, then $E_1 = \{1, 2\}$. The complement of E_1, denoted as E'_1, consists of outcomes 3, 4, 5, 6; that is: $E'_1 = \{3, 4, 5, 6\}$.

PROBLEM

Past records indicate that 25% of the student population drops college algebra. What is the probability that a student does not drop college algebra?

SOLUTION

25% means $\dfrac{25}{100}$ or $\dfrac{1}{4}$ of the students drop algebra.

Those who drop algebra and those who do not drop it constitute the entire universal set.

Apply the formula: $1 - P(E) = P(E')$.

$$1 - \frac{1}{4} = \frac{3}{4}$$

Thus, $\dfrac{3}{4}$ or .75 is the probability that a student does not drop college algebra.

The mathematical formula associated with the probability of a simple event is $P(E) = \dfrac{m}{n}$, where m is the number of favorable outcomes relative to event E, and n is the total number of possible outcomes.

PROBLEM

Given that Jeffrey gets two hits out of every three times at bat at every T-ball game, what is the probability that Jeffrey will get a hit his next time up to bat?

SOLUTION

The number of favorable outcomes (i.e., the number of hits) is 2. The number of possible outcomes is 3. Thus, the probability of a hit is $\frac{2}{3}$.

PROBABILITY OF A OR B

If the probability of one event is not affected by the probability of another, then the probability of either one of them occurring in an experiment is the sum of their individual probabilities; however, one of the instances of both occurring has to be subtracted out to prevent counting twice the probability of both events. The mathematical formula that allows us to find the probability of obtaining either event A or event B is:

$$P(A \text{ or } B) = P(A) + P(B) - P(A \text{ and } B).$$

For example, the probability of selecting a spade or an ace from a deck of cards would be the probability of a spade $\left(\frac{13}{52} = \frac{1}{4}\right)$ plus the probability of an ace $\left(\frac{4}{52} = \frac{1}{13}\right)$, but each of these probabilities includes the ace of spades, so we have to subtract one of the probabilities that the card is the ace of spades $\left(\frac{1}{52}\right)$. Therefore, the formula is:

$$P(\text{spade or ace}) = P(\text{spade}) + P(\text{ace}) - P(\text{ace of spades})$$

$$= \frac{1}{4} + \frac{1}{13} - \frac{1}{52} = \frac{16}{52}.$$

When you think about it, 16 (and not 17) of the cards in a deck are either a spade or an ace or both.

Note: If events A and B are mutually exclusive, $P(A \text{ and } B) = 0$, then the formula above simplifies to $P(A \text{ or } B) = P(A) + P(B)$.

PROBLEM

On a field trip, the teachers counted the orders for a snack and sent the information in with a few people. The orders were for 94 colas and 56 fries. If there were 133 orders, what was the probability of an order for a cola and fries?

SOLUTION

Before using the formula presented above, we must determine $P(\text{cola})$, $P(\text{fries})$, and $P(\text{cola or fries})$.

$$P(\text{cola}) = \frac{94}{133}$$

$$P(\text{fries}) = \frac{56}{133}$$

$$P(\text{cola or fries}) = \frac{133}{133}$$

$$P(A \text{ or } B) = P(A) + P(B) - P(A \text{ and } B)$$

$$\frac{133}{133} = \frac{94}{133} + \frac{56}{133} - P \text{ (cola and fries)}$$

$$\frac{133}{133} = \frac{150}{133} - P \text{ (cola and fries)}$$

$$P(\text{cola and fries}) = \frac{150}{133} - \frac{133}{133}$$

$$P(\text{cola and fries}) = \frac{17}{133}$$

Note that you do not have to actually get a decimal for each fraction, and since the denominators are the same, this reduces to an addition and subtraction problem.

PROBABILITY OF ONE OF TWO EXCLUSIVE EVENTS

Given two events, A and B, if we want the probability that *exactly* one of these occurs, then the formula becomes:

$$P(\text{exactly one of } A \text{ or } B) = P(A) + P(B) - 2 \times P(A \text{ and } B).$$

The factor of 2 in this equation eliminates all possibility of both events taking place, since you want *exactly one* of events A or B to occur, not both. In a previous example, it eliminates the probability of an ace and a spade (the ace of spades) when considering $P(\text{ace})$ plus the probability of a spade and an ace (also the ace of spades) when considering $P(\text{spade})$.

PROBLEM

An ordinary coin is tossed and a fair 6-sided number cube is rolled. What is the probability of getting tails on the coin or getting a 4 on the top face of the number cube, but not both of these events?

SOLUTION

Let A = Event of getting tails on the coin.

Let B = Event of getting a 4 on the number cube.

$$P(A) = \frac{1}{2} \text{ and } P(B) = \frac{1}{6}$$

Then $P(\text{exactly one of } A, B \text{ occurring}) = \frac{1}{2} + \frac{1}{6} - (2)\left(\frac{1}{2}\right)\left(\frac{1}{6}\right) = \frac{1}{2}$

CHECK YOUR WORK!

To verify this result, let H = heads and T = tails. If the coin is tossed first, the twelve possible outcomes for the coin and number cube are written as $H1$, $H2, H3, H4, H5, H6, T1, T2, T3, T4, T5, T6$. The favorable result would consist of six outcomes, namely $T1, T2, T3, T5, T6, H4$. Thus, the corresponding probability is $\frac{6}{12}$ or $\frac{1}{2}$.

CONDITIONAL PROBABILITY

From the **conditional probability** formula:

$$P(A|B) = \frac{P(A \text{ and } B)}{P(B)},$$

where $P(A|B)$ means "the probability of A, given that B has occurred."

We can derive the multiplication rule: $P(A \text{ and } B) = P(A|B) \times P(B)$. It is not necessary to fully understand what conditional probability is as long as you are able to apply the counting rules discussed earlier.

PROBLEM

If Kyle has eight pairs of socks (four white, two black, one blue, and one red). What is the probability that he randomly chooses two pairs of white socks to wear on consecutive days.

SOLUTION

What is the probability of selecting one pair of white socks?

$$\frac{4}{8}$$

Given that Kyle already chose a pair of white socks, what is the probability that he chooses another pair of white socks?

$$\frac{3}{7}$$

This is because he now has 3 pairs of white socks left to choose out of 7 pairs of socks left.

Thus, the probability of drawing two pairs of white socks is:

$$\frac{4}{8} \times \frac{3}{7}, \text{ or } \frac{3}{14}$$

If the occurrence of event A in no way affects the occurrence or non-occurrence of event B, then events A and B are said to be **independent**. If events A and B are independent, then $P(A \text{ and } B) = P(A) \times P(B)$, according to the counting rule.

PROBLEM

The probability that a male child is born is 0.5. What is the probability that the next three unrelated children born at Memorial Hospital are all boys?

SOLUTION

The birth of three unrelated male children in a row would be considered an independent event.

$$P(3 \text{ males}) = (0.5)(0.5)(0.5)$$

$$= 0.125$$

Thus, the probability of three consecutive male unrelated births is 0.125, which is 12.5%.

PROBLEM

An ordinary coin is tossed and a fair 6-sided cube is rolled. What is the probability of getting tails on the coin *and* getting a 4 on the top face of the number cube?

SOLUTION

Since A and B represent independent events, $P(A$ and $B) = P$ (getting tails on the coin and getting a 4 on the number cube) $= P(A) \cdot P(B) = \dfrac{1}{2} \cdot \dfrac{1}{6} = \dfrac{1}{12}$.

PROBABILITY WORD PROBLEMS

As with all prior real-world problems, the context of the problem can vary widely. A table or graph will provide the necessary information to calculate a single outcome, multiple outcomes, conditional probability, or an expected value.

Using Tables for Probability

Tables present data in an easy-to-read format. This table is a breakdown of the student vote for the winner of the election for student body president.

Student Votes for Student Body President

	Freshman	Sophomore	Junior	Senior
Male	9%	17%	10%	9%
Female	16%	13%	15%	11%

PROBLEM

If a student who voted for the winner is selected at random, what is the probability that that person is a sophomore and a male?

SOLUTION

By locating the correct cell in the table, we see that 17% of votes were from sophomore males. This gives a probability of 0.17.

PROBLEM

What is the probability of a randomly selected student not being a senior?

SOLUTION

Locate the senior cells.

9% males + 11% females = 20% seniors

Nonseniors: 100% − 20% = 80%

Thus, the probability of not being a senior is 0.80.

PROBLEM

Knowing that the selected person is a junior, what is the probability that the person is a female?

SOLUTION

This is a conditional probability problem. To solve such a problem, we need to identify how many are in the given category, in this case juniors.

Locate the junior cells: 10% males + 15% females = 25% juniors. The numerical value associated with the given information, that the person is a junior, is the denominator for the fraction.

Of the juniors, identify the percent who are female: 15%. This is the numerator.

Thus, the probability that the person is female, given that the person is a junior, is $\frac{15}{25}$ or 0.60.

Note that you don't have to know *how many* students are juniors or female juniors. Let's say the whole student population is x. Then, according to the table, $.25x$ are juniors and $.15x$ are female juniors, so $P(\text{female}|\text{junior}) = \frac{.15x}{.25x}$. The x's cancel out, and the answer is .60.

If in this election there were 500 votes for the winner, how many females cast votes for the winner?

Total number of females: 16% + 13% + 15% + 11% = 55%

55% of 500 = 275. Thus, 275 females cast votes for the winner.

MEASURES OF CENTRAL TENDENCY

There are three ways to describe the tendency of a set of data, meaning what a "typical" value is: the mean, median, and mode. All three of these numbers are measures of **central tendency**. They describe the "middle" or "center" of the data.

MEAN

The **mean** is the arithmetic average. It is the sum of the variables divided by the total number of variables.

Example:

The mean of 4, 3, and 8 is

$$\frac{4+3+8}{3} = \frac{15}{3} = 5$$

Find the mean hourly wages for four company employees who make $5/hour, $8/hour, $12/hour, and $15/hour.

The mean hourly wage is the average.

$$\frac{\$5 + \$8 + \$12 + \$15}{4} = \frac{\$40}{4} = \$10/\text{hour}$$

MEDIAN

The **median** is the middle value in a set. The set of numbers first needs to be put in order, smallest to largest or vice versa. When there is an odd number of values in the set, the median is simply the middle value, and there is an equal number of values larger and smaller than the median. When the set has an even number of values, the average of the two middle values is the median.

Example:

The median of (2, 3, 5, 8, 9) is 5.

The median of (2, 3, 5, 9, 10, 11) is $\dfrac{5+9}{2} = 7$.

Note that the median doesn't have to be an element of the set. In fact, with an even number of values, it often is not.

Note: The above rules apply, even if some numbers are repeated. For example, the median of (2, 3, 3, 4, 6, 8, 9) is 4.

MODE

The **mode** is the most frequently occurring value in the set of values.

Example:

The mode of (4, 5, 8, 3, 8, 2) would be 8, since it occurs twice whereas the other values occur only once.

There can be two modes (in which case the set is called **bimodal**), or in fact as many modes as you have values.

Example:

The set 2, 2, 3, 3, 5, 5, 8, 8, 9, 9 has five modes, since each value is mentioned twice. If each number in a set of numbers appears only once, there is no mode. *Example*: (2, 3, 5, 8, 9) has no mode. Of course, in this case the mode isn't a very useful measure of central tendency, and the mean or median would describe the data better.

PROBLEM

For this series of observations, find the mean, median, and mode.

500, 600, 800, 800, 900, 900, 900, 900, 900, 1,000, 1,100

SOLUTION

The mean is the value obtained by adding all the measurements and dividing by the number of measurements.

$$\frac{500 + 600 + 800 + 800 + 900 + 900 + 900 + 900 + 900 + 1,000 + 1,100}{11}$$

$$= \frac{9,300}{11} = 845.45$$

The median is the value appearing in the middle. We have 11 values, and they are already in order, so here the sixth, 900, is the median.

The mode is the value that appears most frequently. This is also 900, which appears five times.

COMPARING THE MEAN, MEDIAN, AND MODE IN A VARIETY OF DISTRIBUTIONS

It is sometimes useful to make comparisons about the relative values of the mean, median, and mode. This section presents steps to do that without having to calculate the exact values of these measures.

Step 1: If a bar graph is not provided, sketch one from the information given in the problem.

The graph will either be skewed to the left, skewed to the right, or approximately normal. A **skewed** distribution has one of its tails longer than the other.

- A graph skewed to the left will look like Figure 1; its left tail is longer. The order of the three measures is mean < median < mode (alphabetical order).

- A graph skewed to the right will look like Figure 2; its right tail is longer. The order of the three measures is mode < median < mean (reverse alphabetical order).

- A graph that is approximately **normal** (also called **bell-shaped**) will look like Figure 3, and the mean = median = mode.

Figure 1

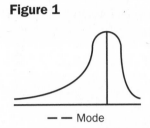

— — Mode

Figure 2

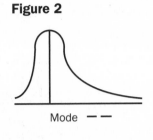

Mode — —

Figure 3

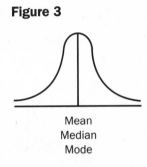

Mean
Median
Mode

Step 2: Write the word "mode" under the highest column of the bar graph, because the mode is the most frequent. If the graph is skewed right or left, the positioning of the mode establishes the order for the remaining two terms according to the information provided above. If the graph is approximately symmetrical, the value of all three terms are approximately equivalent.

PROBLEM

On a trip to the Everglades, students tested the pH of the water at different sites. Most of the pH tests were at 6. A few read 7, and one read 8. Select the statement that is true about the distribution of the pH test results.

A. The mode and the mean are the same.

B. The mode is less than the mean.

C. The median is greater than the mean.

D. The median is less than the mode.

SOLUTION

Sketch a graph.

The graph is skewed to the right. The mode is furthest left. Thus, mode < median < mean.

Choices (A), (C), and (D) do not coincide with what has been established in terms of relative order. (B) is the only choice that does follow from our conclusions. Thus, (B) is the correct response.

Estimate the median of each distribution shown below and describe how the mean compares to it.

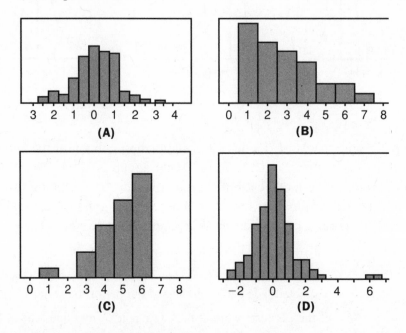

(A)   **(B)**

(C)   **(D)**

Graph (A) is symmetric with a median at about 0. The mean will also be about 0, since it approximates a normal curve. Graph (B) is right-skewed with a median of about 3. The mean will be pulled more toward the values of 4−7 than it will be toward 1−2, so the mean will be greater than 3. Graph (C) is left-skewed with a median of about 5. The mean will be pulled toward the lower values, so it is less than 5. Graph (D) is symmetric with a few extreme high values. The median is about 0, and the mean will be just slightly larger due to the outliers around 6.

MEASURES OF VARIABILITY

In addition to measures of central tendency, distributions need to be described with measures of **variability**, or spread. It is not enough to know where the middle of a distribution is, but also how spread out it is. A manufacturer of light bulbs would like small variability in the amount of hours the bulbs will likely burn. A track coach who needs to decide which athletes go on to the finals may want larger variability in heat times because it will be easier to decide who are truly the fastest runners.

The **range** of a data set is simply the difference between the maximum value and the minimum value.

The range is rarely a good choice to represent the data set, especially because it can be affected by outliers. Look, for example, at graph (D) of the last problem. Its range is -3 to $+7$, but that doesn't show that almost all of the data lie between -2 and $+2$.

For data that are fairly symmetric, the standard deviation and variance are useful measures of variability.

The **variance** tells us how much variability exists in a distribution. It is the "average" of the squared differences between the data values and the mean. The variance is calculated with the formula

$$s^2 = \frac{1}{n-1}\sum(x_i - \bar{x})^2$$

where n is the number of data points, x_i represents each data value, and $\bar{x}$ is the mean. The **standard deviation** is the square root of the variance.

The formula for the standard deviation is therefore

$$s = \sqrt{\frac{1}{n-1}\sum(x_i - \bar{x})^2}$$

The standard deviation is used for most applications in statistics. It can be thought of as the typical distance an observation lies from the mean.

PROBLEM

The average monthly rainfall in inches in Birmingham, England, is shown in the table that follows.

Jan	Feb	Mar	Apr	May	Jun	Jul	Aug	Sep	Oct	Nov	Dec
2.3	1.9	2.1	1.8	2.2	2.2	2.0	2.8	2.2	2.1	2.5	2.6

Compute the variance and standard deviation of the monthly rainfall.

SOLUTION

To compute the variance and standard deviation, we must first compute the mean. In this data set, $\bar{x} = \dfrac{26.7}{12} = 2.225$. The variance is computed as follows:

$$s^2 = \frac{1}{n-1} \sum (x_i - \bar{x})^2$$

$$= \frac{1}{12-1}[(2.3 - 2.225)^2 + (1.9 - 2.225)^2 + (2.1 - 2.225)^2$$

$$+ \ldots + (2.6 - 2.225)^2]$$

$$= \frac{1}{11}[0.9225]$$

$$= 0.084$$

The standard deviation is the square root of the variance, or $\sqrt{0.084} \approx 0.290$. Thus, from this data set, we can say that the monthly mean rainfall in Birmingham is 2.225 inches and it varies by about 0.29 inches from that each month, sometimes more, sometimes less.

DATA ANALYSIS

Data analysis often involves putting numerical values into picture form, such as bar graphs, line graphs, and circle graphs. In this manner, we gain a more intuitive understanding of the given information.

BAR GRAPHS

Bar graphs are used to compare amounts of the same measurements. The following bar graph compares the number of bushels of wheat and corn produced on a farm from 1975 to 1985. The horizontal axis for a bar graph consists of categories (e.g., years, ethnicity, marital status) rather than values, and the widths of the bars are uniform. The emphasis is on the height of the bars. Contrast this with histograms, discussed next.

PROBLEM

According to the graph below, in which year was the least number of bushels of wheat produced?

Number of Bushels (to the Nearest 5 Bushels) of Wheat and Corn Produced by Farm RQS, 1975−1985

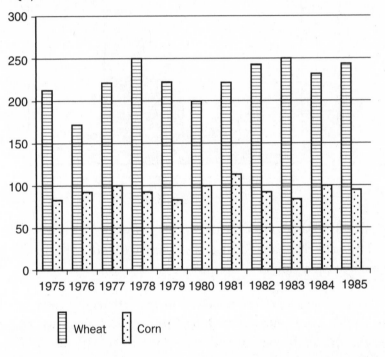

SOLUTION

By inspecting the graph, we find that the shortest bar representing wheat production is the one for 1976. Thus, the least number of bushels of wheat produced in 1975–1985 occurred in 1976.

HISTOGRAMS

A **histogram** is an appropriate display for quantitative data. It is used primarily for continuous data, but may be used for discrete data that have a wide spread. The horizontal axis is broken into intervals that do not have to be of uniform size. Histograms are also good for large data sets. The area of the bar denotes the value, not the height, as in a bar graph.

The histogram below shows the amount of money spent by passengers on a ship during a recent cruise to Alaska.

Example:

Passenger Spending During Cruise to Alaska

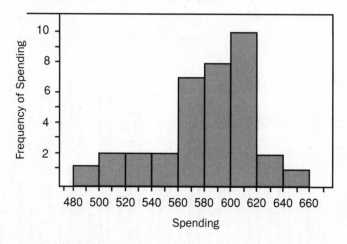

The intervals have widths of $20. One person spent between $480 and $500, two spent between $500 and $520, and so on. We cannot tell from the graph the precise amount each individual spent.

The distribution has a shape skewed to the left with a peak around $600 to $620. The data are centered at about $590—this is about where half of the observations will be to the left and half to the right. The range of the data is about $180, but the clear majority of passengers spent between $560 and $620. There are no extreme values present or gaps within the data.

LINE GRAPHS

Line graphs are very useful in representing data on two different but related subjects. Line graphs are often used to track the changes or shifts in certain factors. In the next problem, a line graph is used to track the changes in the amount of scholarship money awarded to graduating seniors at a particular high school over the span of several years.

PROBLEM

According to the line graph below, by how much did the scholarship money increase between 1987 and 1988?

Amount of Scholarship Money Awarded to Graduating Seniors, West High, 1981–1990

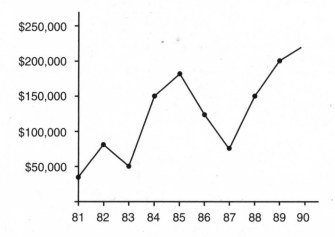

SOLUTION

To find the increase in scholarship money from 1987 to 1988, locate the amounts for 1987 and 1988. In 1987, the amount of scholarship money is half-way between $50,000 and $100,000, or $75,000. In 1988, the amount of scholarship money is $150,000. The increase is thus $150,000 − 75,000 = $75,000.

PIE CHARTS

Circle graphs (or **pie charts**) are used to show the breakdown of a whole picture. When the circle graph is used to demonstrate this breakdown in terms of percents, the whole figure represents 100% and the parts of the circle graph represent percentages of the total. When added together, these percentages add up to 100%. The circle graph in the next problem shows how a family's budget has been divided into different categories by using percentages.

PROBLEM

Using the budget shown below, a family with an income of $3,000 a month would plan to spend what amount on housing?

Family Budget

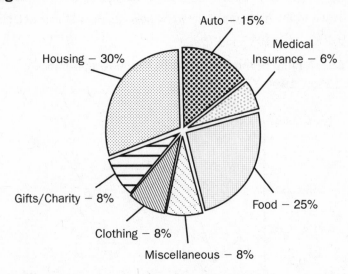

To find the amount spent on housing, locate on the pie chart the percentage allotted to housing, or 30%. Then calculate 30% of $3,000 = $900. The family plans to spend $900 on housing.

STEMPLOTS

A **stemplot**, also called stem-and-leaf plot, can be used to display univariate data as well. It is good for small sets of data (about 50 or less) and forms a plot much like a histogram. This stemplot represents test scores for a class of 32 students.

Test Scores

3	3
4	
5	
6	3 7 9
7	2 2 5 7
8	1 2 6 8 8 8 9 9 9
9	0 0 0 1 3 3 4 5 5 6 7
10	0 0 0 0
Key: 6	3 represents a score of 63

The values on the left of the vertical bar are called the stems; those on the right are called leaves. Stems and leaves need not be tens and ones—they may be hundreds and tens, ones and tenths, and so on. A good stemplot always includes a key for the reader so that the values may be interpreted correctly.

PROBLEM

Describe the distribution of test scores for students in the class using the stemplot.

SOLUTION

The distribution of the test scores is skewed toward lower values (to the left). It is centered at about 89 with a range of 67. There is an extreme low value at 33, which appears to be an outlier. Without it, the range is only 37, about half as much. The test scores have a mean of approximately 85.4, a median of 89, and a mode of 100.

ANALYZING PATTERNS IN SCATTERPLOTS

Bivariate data consist of two variables. Typically, we are looking for an association between these two variables. The variables may be categorical or quantitative; in this section, we focus on quantitative bivariate data. **Scatterplots** are used to visualize quantitative bivariate data.

The two variables under study are referred to as the **explanatory variable** (x) and the **response variable** (y). The explanatory variable *explains* or *predicts* the response variable. The response variable measures the outcomes that have been observed.

PROBLEM

Data collected from the labels on snack foods included the number of grams of fat per serving and the total number of calories in the food. Identify which are the explanatory and response variables when looking for a relationship between fat grams and calories.

SOLUTION

The explanatory variable is grams of fat and the response variable is calories. The number of grams of fat would be a predictor of the number of calories in the snack.

Scatterplots can tell us if and how two variables are related. When we examined univariate data in the preceding sections, we described a distribution's shape, center, spread, and outliers/unusual features. In a scatterplot, we focus on its shape, direction, and strength, and we look for outliers and unusual features. Below is a scatterplot of the top 30 leading scorers in the history of the National Basketball Association (NBA). Each point represents 1 of the 30 players. Michael Jordan, who scored 32,292 points in 1,072 games, is noted.

NBA Top 30 Scorers

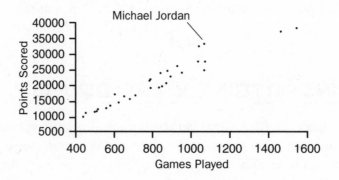

The **shape** of a plot is usually classified as linear or nonlinear (curved). The **direction** of a scatterplot tells what happens to the response variables as the explanatory variable increases. This is the slope of the general pattern of the data. The **strength** describes how tight or spread out the points of a scatterplot are.

The three scatterplots below show comparisons of various directions of the data. The top two have a clear linear trend, whereas the third scatterplot shows a random type of distribution with no clear association among points.

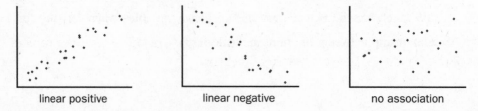

| linear positive | linear negative | no association |

When analyzing a scatterplot, it is also a good idea to look for outliers, clusters, or gaps in the data. The scatterplot below has an obvious gap. There is an overall positive, linear association, but we should find out the reason for the gap.

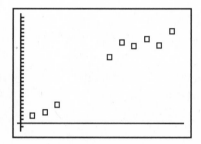

The scatterplot below has an obvious outlier. An outlier falls outside the general pattern of the data. There could be several possible reasons for the outlier, which merits investigation.

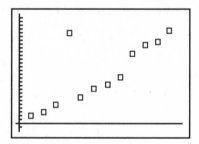

PROBLEM

In the scatterplot of the top 30 scorers in NBA history shown at the beginning of this section, identify the explanatory variable and the response variable. Describe the association between the two variables.

SOLUTION

The explanatory variable is the number of games played. The response variable is the number of points scored. The relationship is linear and positive. There are no outliers, but there is a large gap between 1,100 and 1,400 games. The two extreme points, probably a couple of players with very long careers, appear to follow the general pattern of the data.

Drill Questions

1. What is the value of the median in the following sample?

   {19, 15, 21, 24, 11}

   (A) 18
   (B) 19
   (C) 20
   (D) 21

2. The sample standard deviation, s, of a group of data is given by the formula

   $s = \sqrt{\dfrac{\sum_{i=1}^{n}(x_i - \bar{x})^2}{n-1}}$, where x_i represents each data point, $\bar{x}$ represents the

   mean, and n represents the number of data points. What is the sample standard deviation for the following data (rounded off to the nearest hundredth)?

   {12, 15, 20, 21}

   (A) 3.67
   (B) 3.81
   (C) 4.24
   (D) 8.25

3. Mrs. Smith teaches a class of 150 students. On a recent exam administered to her class, 100 students scored 90 or better, 40 students scored between 80 and 89, and the remaining students scored between 70 and 79. Which one of the following statements is correct concerning the students' exam scores?

   (A) The median equals the mean.
   (B) The mean is less than the mode.
   (C) The median is less than the mean.
   (D) The median is greater than the mode.

4. A jar contains 20 balls, numbered 1 through 20. What is the probability that a randomly chosen ball has a number on it that is divisible by 4?

    (A) $\dfrac{1}{20}$

    (B) $\dfrac{1}{10}$

    (C) $\dfrac{1}{5}$

    (D) $\dfrac{1}{4}$

5. From four textbooks of college math and three textbooks of college physics, a researcher will choose two math textbooks and one physics textbook. How many different groupings of three textbooks are there?

    (A) 12
    (B) 18
    (C) 24
    (D) 36

6. Four out of five dentists surveyed recommended professional teeth cleaning twice a year. Of the dentists who made this recommendation, 10% said they are underpaid. If a dentist from this survey is randomly selected, what is the probability that this selected dentist has recommended professional teeth cleaning twice a year and also said that he or she is underpaid?

    (A) $\dfrac{2}{25}$

    (B) $\dfrac{4}{15}$

    (C) $\dfrac{2}{5}$

    (D) $\dfrac{7}{10}$

7. An "unusual" number cube has five faces, numbered 1 through 5. This number cube will be rolled twice. Assuming that each of the five numbers has the same likelihood of appearing, what is the probability that *at least one* of the following events will occur?

    Event 1: The number 3 will appear on the first roll.

    Event 2: An even number will appear on the second roll.

    (A) $\dfrac{4}{5}$

    (B) $\dfrac{3}{5}$

    (C) $\dfrac{13}{25}$

    (D) $\dfrac{6}{25}$

8. In a class of 30 students, 24 are female. There are a total of 12 history majors, and 4 of these history majors are male. How many non-history female students are there?

    (A) 16
    (B) 14
    (C) 12
    (D) 10

9. The mean weight of a group of ten people in a room is 160 pounds. When two people leave, the mean weight of the remaining eight people is 166 pounds. What is the mean weight, in pounds, of the two people who leave the room?

    (A) 136
    (B) 145
    (C) 154
    (D) 163

10. At the MNP company, each person gets an identification number. The number contains five digits, chosen from the digits 1 through 8, inclusive. The left-most digit must be 2. If repetition of any digit is allowed, how many different identification numbers are possible?

    (A) 32,768
    (B) 8192
    (C) 6720
    (D) 4096

Answers to Drill Questions

1. **(B)** The median of five numbers is the third number, when the numbers are arranged in ascending order. Rewrite the given sample as {11, 15, 19, 21, 24}. Then the median is the third number, which is 19.

2. **(C)** First, we need to find the value of $\overline{X}$, which is the mean. The value of $\overline{X}$ is given by $\dfrac{12+15+20+21}{4} = \dfrac{68}{4} = 17$. Next, substitute each data value into the given formula for s. Then $s = \sqrt{\dfrac{(12-17)^2 + (15-17)^2 + (20-17)^2 + (21-17)^2}{4-1}} = \sqrt{\dfrac{(-5)^2 + (-2)^2 + 3^2 + 4^2}{3}} = \sqrt{\dfrac{25+4+9+16}{3}} = \sqrt{\dfrac{54}{3}} = \sqrt{18} \approx 4.24$.

3. **(B)** There were only $150 - 100 - 40 = 10$ students who scored in the lowest of the three grade groups. Note that the largest number of students scored in the highest of the grade groups. By definition, the graph of this distribution must skewed to the left. Thus, the mean is less than the median and the median is less than the mode. Therefore, the mean is less than the mode.

4. **(D)** There are five balls whose associated number is divisible by 4, namely those numbered 4, 8, 12, 16, and 20. The probability of randomly choosing one of these balls is $\dfrac{5}{20}$, which reduces to $\dfrac{1}{4}$.

5. **(B)** The number of ways to choose a group of two out of four math textbooks is given by the formula $_4C_2 = \dfrac{4!}{2!(4-2)!} = \dfrac{24}{(2)(2)} = \dfrac{24}{4} = 6$. In

addition, there are only 3 ways to select one out of three physics textbooks. Thus, the number of different groupings for choosing the two math books and one physics book is $(6)(3) = 18$.

6. **(A)** Let A represent the event that a dentist recommended professional teeth cleaning twice a year. Let B represent the event that a dentist said that he or she is underpaid. Then $P(B \mid A)$ represents the probability that a dentist said that he or she is underpaid, given that this dentist recommended professional teeth cleaning twice a year. We will use the formula $P(B \mid A) = \dfrac{P(B \text{ and } A)}{P(A)}$, where $P(A) = \dfrac{4}{5} = 0.80$ and $P(B \mid A) = 10\% = 0.10$. By substitution, $0.10 = \dfrac{P(B \text{ and } A)}{0.80}$. Thus, $P(B \text{ and } A) = (0.10)(0.80) = 0.08 = \dfrac{2}{25}$.

7. **(C)** Let A represent the event that the number 3 appears on the first roll. Let B represent the event that an even number appears on the second roll. Out of the five numbers, only one is numbered 3 and there are two even numbers (2 and 4). Then $P(A) = \dfrac{1}{5}$ and $P(B) = \dfrac{2}{5}$. We will use the formula $P(A \text{ or } B) = P(A) + P(B) - P(A \text{ and } B)$, where $P(A \text{ or } B)$ represents the probability that at least one of A or B will occur. Note that these two events are independent because the occurrence of one of them has no effect on the probability of the occurrence of the other. This means that $P(A \text{ and } B) = P(A) \times P(B)$. Now, by substitution, $P(A \text{ or } B) = \dfrac{1}{5} + \dfrac{2}{5} - \left(\dfrac{1}{5}\right)\left(\dfrac{2}{5}\right) = \dfrac{3}{5} - \dfrac{2}{25} = \dfrac{15}{25} - \dfrac{2}{25} = \dfrac{13}{25}$.

8. **(A)** Since there are 4 male history majors, there must be $12 - 4 = 8$ female history majors. There are are total of 24 female students, so the number of female students who are non-history majors is $24 - 8 = 16$.

9. **(A)** The total weight for all ten people is $(160)(10) = 1{,}600$ pounds. After two people leave the room, the total weight for the remaining eight people is $(166)(8) = 1{,}328$ pounds. This means that the combined weight of the two people who left the room is $1{,}600 - 1{,}328 = 272$ pounds. Thus, the mean weight of these two people is $\dfrac{272}{2} = 136$ pounds.

10. **(D)** Since only the number 2 is allowed for the left-most digit, there is only one choice for this digit. For each of the other four digits, there are no restrictions on the digits 1 through 8. This means that there are eight choices for each of these four digits. Thus, there are $(1)(8)(8)(8)(8) = 4{,}096$ different identification numbers.

CHAPTER 8

Logic

LOGIC

The topic of logic encompasses a wealth of subjects related to the principles of reasoning. Here, we will be concerned with logic on an elementary level, and you will recognize the principles introduced here because not only have you been using them in your everyday life, you also have seen them in one form or another in this book. The chapter is designed to familiarize you with the terminology you might be expected to know as well as thought processes that you can use in everyday life.

Sentential calculus is the "calculus of sentences," a field in which the truth or falseness of assertions is examined by using algebraic tools. We will approach logic from a "true" or "false" perspective here. This chapter contains many examples of sentences to illustrate the terms that are defined.

SENTENCES

A **sentence** is any expression that can be labeled either true or false.

Examples:

Expressions to which the terms "true" or "false" can be assigned include the following:

1. "It is raining where I am standing."

2. "My name is George."

3. "1 + 2 = 3"

Examples:

Expressions to which the terms "true" or "false" cannot be assigned include the following:

1. "I will probably be healthier if I exercise."

2. "It will rain on this day, one year from now."

3. "What I am saying at this instant is a lie."

Sentences can be combined to form new sentences using the connectives **AND, OR, NOT,** and **IF-THEN**.

Examples:

The sentences

1. "John is tired."

2. "Mary is cooking."

can be combined to form

1. "John is tired AND Mary is cooking."

2. "John is tired OR Mary is cooking."

3. "John is NOT tired."

4. "IF John is tired, THEN Mary is cooking."

LOGICAL PROPERTIES OF SENTENCES

Consistency

A sentence is **consistent** if and only if it is *possible* that it is true. A sentence is **inconsistent** if and only if it is not consistent; that is, if and only if it is *impossible* that it is true.

Example:

"At least one odd number is not odd" is an inconsistent sentence.

Logical Truth

A sentence is **logically true** if and only if it is *impossible* for it to be false; that is, the denial of the sentence is inconsistent.

Example:

Either Mars is a planet or Mars is not a planet.

Logical Falsity

A sentence is **logically false** if and only if it is *impossible* for it to be true; that is, the sentence is inconsistent.

Example:

Mars is a planet and Mars is not a planet.

Logical Indeterminacy (Contingency)

A sentence is **logically indeterminate** (contingent) if and only if it is neither logically true nor logically false.

Example:

Einstein was a physicist and Pauling was a chemist.

Logical Equivalent of Sentences

Two sentences are **logically equivalent** if and only if it is *impossible* for one of the sentences to be true while the other sentence is false; that is, if and only if it is impossible for the two sentences to have different truth values.

Example:

"Chicago is in Illinois and Pittsburgh is in Pennsylvania" is logically equivalent to "Pittsburgh is in Pennsylvania and Chicago is in Illinois."

STATEMENTS

A **statement** is a sentence that is either true or false, but not both.

The following terms and their definitions should become familiar to you. Their logic is probably familiar, even though you haven't as yet given it a label.

CONJUNCTION

If a and b are statements, then a statement of the form "a and b" is called the **conjunction** of a and b, denoted by $a \wedge b$.

DISJUNCTION

The **disjunction** of two statements a and b is shown by the compound statement "a or b," denoted by $a \vee b$.

NEGATION

The **negation** of a statement q is the statement "not q," denoted by $\sim q$.

IMPLICATION

The compound statement "if a, then b," denoted by $a \rightarrow b$, is called a **conditional statement** or an **implication**. "If a" is called the **hypothesis** or **premise** of the implication, and "then b" is called the **conclusion** of the implication. Further, statement a is called the **antecedent** of the implication, and statement b is called the **consequent** of the implication.

CONVERSE

The **converse** of $a \rightarrow b$ is $b \rightarrow a$.

CONTRAPOSITIVE

The **contrapositive** of $a \rightarrow b$ is $\sim b \rightarrow \sim a$.

INVERSE

The **inverse** of $a \rightarrow b$ is $\sim a \rightarrow \sim b$.

BICONDITIONAL

The statement of the form "p if and only if q," denoted by $p \leftrightarrow q$, is called a **biconditional** statement.

VALIDITY

An argument is **valid** if the truth of the premises means that the conclusions must also be true.

INTUITION

Intuition is the process of making generalizations on insight.

PROBLEM

Write the inverse for each of the following statements. Determine whether the inverse is true or false.

1. If a person is stealing, he is breaking the law.

2. If a line is perpendicular to a segment at its midpoint, it is the perpendicular bisector of the segment.

3. Dead men tell no tales.

SOLUTION

The inverse of a given conditional statement is formed by negating both the hypothesis and conclusion of the conditional statement.

1. The hypothesis of this statement is "a person is stealing"; the conclusion is "he is breaking the law." The negation of the hypothesis is "a person is not stealing." The inverse is "if a person is not stealing, he is not breaking the law."

   The inverse is false, since there are more ways to break the law than by stealing. Clearly, a murderer may not be stealing but he is surely breaking the law.

2. In this statement, the hypothesis contains two conditions: a) the line is perpendicular to the segment; and b) the line intersects the segment at the midpoint. The negation of (statement a *and* statement b) is (not statement a *or* not statement b). Thus, the negation of the hypothesis is "The line is not perpendicular to the segment or it doesn't intersect the segment at the midpoint." The negation of the conclusion is "the line is not the perpendicular bisector of a segment."

   The inverse is "if a line is not perpendicular to the segment or does not intersect the segment at the midpoint, then the line is not the perpendicular bisector of the segment."

   In this case, the inverse is true. If either of the conditions holds (the line is not perpendicular; the line does not intersect at the midpoint), then the line cannot be a perpendicular bisector.

3. This statement is not written in if-then form, which makes its hypothesis and conclusion more difficult to see. The hypothesis is implied to be "the man is dead"; the conclusion is implied to be "the man tells no tales." The inverse is, therefore, "If a man is not dead, then he will tell tales."

   The inverse is false. Many witnesses to crimes are still alive but they have never told their stories to the police, probably out of fear or because they didn't want to get involved.

BASIC PRINCIPLES, LAWS, AND THEOREMS

1. Any statement is either true or false. (The Law of the Excluded Middle)

2. A statement cannot be both true and false. (The Law of Contradiction)

3. The converse of a true statement is not necessarily true.

4. The converse of a definition is always true.

5. For a theorem to be true, it must be true for all cases.

6. A statement is false if one false instance of the statement exists.

7. The inverse of a true statement is not necessarily true.

8. The contrapositive of a true statement is true and the contrapositive of a false statement is false.

9. If the converse of a true statement is true, then the inverse is true. Likewise, if the converse is false, the inverse is false.

10. Statements that are either both true or both false are said to be **logically equivalent**.

NECESSARY AND SUFFICIENT CONDITIONS

Let P and Q represent statements. "If P, then Q" is a conditional statement in which P is a sufficient condition for Q, and similarly Q is a necessary condition for P.

Example:

Consider the statement: "If it rains, then Jane will go to the movies." "If it rains" is a sufficient condition for Jane to go to the movies. "Jane will go to the movies" is a necessary condition for rain to have occurred.

Note that for the statement given, "If it rains" may not be the only condition for which Jane goes to the movies; however, it is a *sufficient* condition. Likewise, "Jane will go to the movies" will certainly not be the only result from a rainy weather condition (for example, "the ground will get wet" is another likely conclusion). However, knowing that Jane went to the movies is a *necessary* condition for rain to have occurred.

In the biconditional statement "P if and only if Q," P is a necessary and sufficient condition for Q, and vice versa.

Example:

Consider the statement "Rick gets paid if and only if he works." "Rick gets paid" is both a sufficient and necessary condition for him to work. Also, Rick's working is a sufficient and necessary condition for him to get paid.

Thus, we have the following basic principles to add to our list of ten from the preceding section:

11. If a given statement and its converse are both true, then the conditions in the hypothesis of the statement are both necessary and sufficient for the conclusion of the statement.

12. If a given statement is true but its converse is false, then the conditions are sufficient but not necessary for the conclusion of the statement.

13. If a given statement and its converse are both false, then the conditions are neither sufficient nor necessary for the statement's conclusion.

DEDUCTIVE REASONING

An arrangement of statements that would allow you to deduce the third one from the preceding two is called a **syllogism**. A syllogism has three parts:

1. The first part is a general statement concerning a whole group. This is called the **major premise**.

2. The second part is a specific statement which indicates that a certain individual is a member of that group. This is called the **minor premise**.

3. The last part of a syllogism is a statement to the effect that the general statement which applies to the group also applies to the individual. This third statement of a syllogism is called a **deduction**.

Example:

This is an example of a properly deduced argument.

A. Major Premise: All birds have feathers.

B. Minor Premise: An eagle is a bird.

C. Deduction: An eagle has feathers.

The technique of employing a syllogism to arrive at a conclusion is called **deductive reasoning**.

If a major premise that is true is followed by an appropriate minor premise that is true, a conclusion can be deduced that must be true, and the reasoning is valid. However, if a major premise that is true is followed by an *inappropriate* minor premise that is also true, a conclusion cannot be deduced.

Example:

This is an example of an improperly deduced argument.

A. Major Premise: All people who vote are at least 18 years old.

B. Improper Minor Premise: Jane is at least 18.

C. Illogical Deduction: Jane votes.

The flaw in this example is that the major premise in statement A makes a condition on people who vote, not on a person's age. If statements B and C are interchanged, the resulting three-part deduction would be logical.

In the following we will use capital letters X, Y, Z, ... to represent sentences, and develop algebraic tools to represent new sentences formed by linking them with the above connectives. Our connectives may be regarded as operations transforming one or more sentences into a new sentence. To describe them in greater detail, we introduce symbols to represent them. You will find that different symbols representing the same idea may appear in different references.

TRUTH TABLES AND BASIC LOGICAL OPERATIONS

The **truth table** for a sentence X is the exhaustive list of possible logical values of X.

The **logical value** of a sentence X is true (or T) if X is true, and false (or F) if X is false.

NEGATION

If X is a sentence, then $\sim X$ represents the **negation**, the opposite, or the contradiction of X. Thus, the logical values of $\sim X$ are as shown in Table 8-1, where $\sim$ is called the **negation operation** on sentences.

Table 8-1   Truth Table for Negation

X	~X
T	F
F	T

Example:

For X = "Jane is eating an apple," we have

$\sim X$ = "Jane is *not* eating an apple."

The negation operation is called *unary*, transforming a sentence into a unique image sentence.

IFF

We use the symbol **IFF** to represent the expression "if and only if."

AND

For sentences X and Y, the conjunction "X AND Y," represented by $X \wedge Y$, is the sentence that is true IFF both X and Y are true. The truth table for $\wedge$ (or AND) is shown in Table 8-2, where $\wedge$ is called the **conjunction operator**.

Table 8-2   Truth Table for AND

X	Y	X∧Y
T	T	T
T	F	F
F	T	F
F	F	F

The conjunction $\wedge$ is a *binary* operation, transforming a pair of sentences into a unique image sentence.

Example:

For X = "Jane is eating an apple" and Y = "All apples are sweet," we have $X \wedge Y$ = "Jane is eating an apple AND all apples are sweet."

AND/OR

For sentences X and Y, the disjunction "X AND/OR Y," represented by $X \vee Y$, denotes the sentence that is true if either or both X and Y are true. The truth table for $\vee$ is shown in Table 8-3, where $\vee$ is called the **disjunction operator**.

Table 8-3   Truth Table for AND/OR

X	Y	X ∨ Y
T	T	T
T	F	T
F	T	T
F	F	F

As with the conjunction operator, the disjunction is a *binary* operation, transforming the pair of sentences X, Y into a unique image sentence $X \vee Y$.

Example:

For X = "Jane is eating the apple" and Y = "Marvin is running," we have $X \vee Y$ = "Jane is eating the apple AND/OR Marvin is running."

IF-THEN

For sentences X and Y, the **implication** $X \to Y$ represents the statement "IF X THEN Y." $X \to Y$ is false IFF X is true and Y is false; otherwise, it is true. The truth table for $\to$ is shown in Table 8-4. $\to$ is referred to as the **implication operator**.

Table 8-4   Truth Table for IF-THEN

X	Y	$X \rightarrow Y$
T	T	T
T	F	F
F	T	T
F	F	T

Implication is a *binary* operation, transforming the pair of sentences X and Y into a unique image sentence $X \rightarrow Y$.

LOGICAL EQUIVALENCE

For sentences X and Y, the **logical equivalence** $X \leftrightarrow Y$ is true IFF X and Y have the same truth value; otherwise, it is false. The truth table for $\leftrightarrow$ is shown by Table 8-5, where $\leftrightarrow$ represents logical equivalence, "IFF."

Table 8-5   Truth Table for Equivalence

X	Y	$X \leftrightarrow Y$
T	T	T
T	F	F
F	T	F
F	F	T

Example:

For $X =$ "Jane eats apples" and $Y =$ "apples are sweet," we have $X \leftrightarrow Y =$ "Jane eats apples IFF apples are sweet."

Equivalence is a *binary* operation, transforming pairs of sentences X and Y into a unique image sentence $X \leftrightarrow Y$. The two sentences X, Y for which $X \leftrightarrow Y$ are said to be logically equivalent.

LOGICAL EQUIVALENCE VERSUS "MEANING THE SAME"

Logical equivalence ($\leftrightarrow$) is not the same as an equivalence of meanings. Thus, if Jane is eating an apple and Barbara is frightened of mice, then for $X =$ "Jane is eating an apple" and $Y =$ "Barbara is frightened of mice," X and Y are logically equivalent, since both are correct. However, they do not have the same meaning. Statements having the same meaning are, for example, the double negative $\sim\sim X$ (not-not) and X itself.

THEOREM 1—Double Negation Equals Identity

For any sentence X,

$\sim\sim X \leftrightarrow X.$

FUNDAMENTAL PROPERTIES OF OPERATIONS

The next three theorems will look familiar because they were introduced in another format in Chapter 2 for sets and Chapter 3 for the real number system.

THEOREM 2—Properties of Conjunction Operation

For any sentences X, Y, Z, the following properties hold:

1. Commutativity: $X \wedge Y \leftrightarrow Y \wedge X$

2. Associativity: $X \wedge (Y \wedge Z) \leftrightarrow (X \wedge Y) \wedge Z$

THEOREM 3—Properties of Disjunction Operation

For any sentences X, Y, Z, the following properties hold:

1. Commutativity: $X \vee Y \leftrightarrow Y \vee X$

2. Associativity: $X \vee (Y \vee Z) \leftrightarrow (X \vee Y) \vee Z$

THEOREM 4—Distributive Laws

For any sentences X, Y, Z, the following laws hold:

1. $X \vee (Y \wedge Z) \leftrightarrow (X \vee Y) \wedge (X \vee Z)$

2. $X \wedge (Y \vee Z) \leftrightarrow (X \wedge Y) \vee (X \wedge Z)$

THEOREM 5—De Morgan's Laws for Sentences

For any sentences X, Y, the following laws hold:

1. $\sim(X \wedge Y) \leftrightarrow (\sim X) \vee (\sim Y)$

2. $\sim(X \vee Y) \leftrightarrow (\sim X) \wedge (\sim Y)$

Proof of Part 1 of Theorem 5

We can prove $\sim(X \wedge Y) \leftrightarrow (\sim X) \vee (\sim Y)$ by developing a truth table over all possible combinations of X and Y and observing that all values assumed by the sentences are the same. To this end, we first evaluate the expression $\sim(X \wedge Y)$ in Table 8-6a.

Table 8-6a   Truth Table for Negation of Conjunction

X	Y	X∧Y	~(X∧Y)
T	T	T	F
T	F	F	T
F	T	F	T
F	F	F	T

Now we evaluate $(\sim X) \vee (\sim Y)$ in Table 8-6b.

Table 8-6b   Truth Table for Disjunction of Negation

X	Y	~X	~Y	(~X)∨(~Y)
T	T	F	F	F
T	F	F	T	T
F	T	T	F	T
F	F	T	T	T

The last columns of the truth tables coincide, proving our assertion.

THEOREM 6—Two Logical Identities

For any sentences X, Y, the sentences X and $(X \wedge Y) \vee (X \wedge \sim Y)$ are logically equivalent. That is,

$$(X \wedge Y) \vee (X \wedge \sim Y) \leftrightarrow X$$

This is proven in Table 8-7a.

Table 8-7a   Truth Table for (X ∧ Y) ∨ (X ∧ ~Y) ↔ X

X	Y	~Y	X∧Y	X∧~Y	(X∧Y)∨(X∧~Y)
T	T	F	T	F	T
T	F	T	F	T	T
F	T	F	F	F	F
F	F	T	F	F	F

For any sentences X, Y, the sentences X and $X \lor (Y \land \sim Y)$ are logically equivalent. That is

$$X \lor (Y \land \sim Y) \leftrightarrow X$$

This is proven in Table 8-7b.

Table 8-7b   Truth Table for X ∨ (Y ∧ ~Y) ↔ X

X	Y	~Y	Y∧~Y	X∨(Y∧~Y)
T	T	F	F	T
T	F	T	F	T
F	T	F	F	F
F	F	T	F	F

For any sentences X, Y, $(X \rightarrow Y)$ and $(\sim X \lor Y)$ are logically equivalent.

This is proven in Table 8-7c.

Table 8-7c

X	Y	X → Y	~X	~X ∨ Y
T	T	T	F	T
T	F	F	F	F
F	T	T	T	T
F	F	T	T	T

THEOREM 7—Proof by Contradiction

For any sentences X, Y, the following holds:

$$X \rightarrow Y \leftrightarrow {\sim}Y \rightarrow {\sim}X$$

To prove this, we consider Table 8-8.

Table 8-8 Truth Table for Proof by Contradiction

X	Y	X → Y	~Y	~X	~Y → ~X
T	T	T	F	F	T
T	F	F	T	F	F
F	T	T	F	T	T
F	F	T	T	T	T

SENTENCES, LITERALS, AND FUNDAMENTAL CONJUNCTIONS

We have seen that logically equivalent sentences may be expressed in different ways, the simplest examples being that a sentence is equal to its double negation,

$$\sim\sim X \leftrightarrow X,$$

and by De Morgan's theorem,

$$X \vee Y \leftrightarrow {\sim}({\sim}X \wedge {\sim}Y)$$

The significance of sentential calculus and the algebra of logic is that it provides us with a method of producing a "standard" form for representing a statement in terms of the literals. This is indeed unique and, although usually the simplest representation, it does serve as a standard form for comparison and evaluation of sentences.

Drill Questions

1. If P and Q represent statements, which one of the following is equivalent to "Not P and not Q"?

    (A) Not P or not Q
    (B) Not P or Q
    (C) Not $(P$ or $Q)$
    (D) Not $(P$ and $Q)$

2. Let R and S represent statements. Consider the following:

    I.   If R then S
    II.   Not R and S
    III. If S then R

    Which of the above statements is (are) equivalent to the statement "R is a necessary condition for S"?

    (A) Only I
    (B) I and II
    (C) II and III
    (D) Only III

3. What is the inverse of the statement "If it is snowing, then people stay indoors"?

    (A) If people stay indoors, then it is snowing.
    (B) If it is not snowing, then people do not stay indoors.
    (C) If people do not stay indoors, then it is not snowing.
    (D) If it is not snowing, then people stay indoors.

4. What is the negation for the statement "Image is important or personality matters"?

   (A) Image is important and personality does not matter.
   (B) Image is not important or personality does not matter.
   (C) Image is important or personality does not matter.
   (D) Image is not important and personality does not matter.

5. Given any two statements P and Q, where Q is a false statement, which one of the following *must* be false?

   (A) Not P and Q
   (B) Not (P and Q)
   (C) P or not Q
   (D) P implies Q

6. What is the contrapositive of the statement "If Johnny is lying, then Marie will be angry"?

   (A) If Johnny is not lying, then Marie will be angry.
   (B) If Marie is not angry, then Johnny is not lying.
   (C) If Johnny is not lying, then Marie will not be angry.
   (D) If Marie is angry, then Johnny is lying.

7. Which of the following is an example of a biconditional statement?

   (A) A triangle has an angle less than 90° if it is acute.
   (B) A rectangle is a square if and only if it has four equal sides.
   (C) A circle has both a radius and a diameter.
   (D) A cube has either six edges or it has more than six edges.

8. "All of P is in Q and some of R is in P." Based on the previous statement, which one of the following is a valid conclusion?

   (A) Some of R is not in Q.
   (B) All of R is in Q.
   (C) Some of R is in Q.
   (D) None of R is in Q.

9. Given three statements, P, Q, and R, suppose it is known that R is true. Which one of the following must be true?
   (A) $(P \wedge Q) \to R$
   (B) $R \to (P \vee Q)$
   (C) $(P \to Q) \wedge R$
   (D) $(R \to Q) \vee P$

10. For which one of the following statements is the converse true?

   (A) If you live in Toledo, then you live in Ohio.
   (B) If your pet is a cat, then it has a tail.
   (C) If a number is negative, then its square is positive.
   (D) If a geometric figure has three sides, then it is a triangle.

Answers to Drill Questions

1. **(C)** One of De Morgan's Laws for sentences is $\sim(X \vee Y) \leftrightarrow -X \wedge \sim Y$. Substituting P for X and Q for Y, we have $\sim(P \vee Q) \leftrightarrow \sim P \wedge \sim Q$. This statement is read as follows: "Not (P or Q)" is equivalent to "Not P and not Q."

2. **(D)** If R is a necessary condition for S, then by definition S implies R. Also, the statement that "S implies R" is equivalent to the statement "If S then R," which represents item III. Neither of items I or II is equivalent to the given statement.

3. **(B)** The inverse of "If P then Q" is "If not P then not Q." Let P represent the statement "It is snowing" and let Q represent the statement "People stay indoors." By substitution, for the statement "If it is snowing, then people stay indoors," the inverse statement is "If it is not snowing, then people do not stay indoors."

4. **(D)** The negation of $(P \vee Q)$ is $\sim(P \vee Q)$. By one of De Morgan's Laws, the statement $\sim(P \vee Q)$ is equivalent to $\sim P \wedge \sim Q$. Let P represent the statement "Image is important" and let Q represent the statement "Personality matters." Then the negation for "Image is important or personality matters" is "Not (Image is important or personality matters)." This latter statement is equivalent to "Image is not important and personality does not matter."

5. **(A)** Given that Q is a false statement, the statement "Not P and Q" must be false. Any compound statement with the conjunction operator (which is "and") is false unless both component parts are true.

6. **(B)** The contrapositive of "If P then Q" is "If not Q then not P." Let P represent the statement "Johnny is lying" and let Q represent the statement "Marie will be angry." By substitution, for the statement "If Johnny is lying, then Marie will be angry," the contrapositive statement is "If Marie is not angry, then Johnny is not lying."

7. **(B)** A statement is called biconditional if it can be written in the form "P if and only if Q." Let P represent the statement "A rectangle is a square" and let Q represent the statement "It has four equal sides." Then the statement "A rectangle is a square if and only if it has four equal sides" is biconditional. Note that a biconditional statement may also appear in the form "If P then Q, and if Q then P."

8. **(C)** Here are the three possible diagrams for sets P, Q, and R.

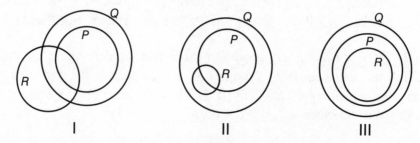

I   II   III

The statement "Some of R is in Q" is true for each of diagrams I, II, and III. Choice (A) is incorrect because it is false for diagrams II and III. Choice (B) is incorrect because it is false for diagram I. Choice (D) is incorrect because it is false for all three diagrams. Note that the statement "Some of R is in P" is true even if all of R is actually in P. The word "some" means "at least one."

9. **(A)** A conditional statement "$X \rightarrow Y$" is true in all instances except when X is true and Y is false. Let X be represented by $P \wedge Q$ and let Y be represented by R. Since R is known to be true and we do not know the truth value of "$P \wedge Q$," we have either "True $\rightarrow$ True" or "False $\rightarrow$ True." In either case "$(P \wedge Q) \rightarrow R$" must be true.

10. **(D)**   The converse of "If X then Y" is "If Y then X." Thus, the converse of choice (D) is "If a geometric figure is a triangle, then it has three sides." This is a true statement. Choice (A) is incorrect because its converse "If you live in Ohio, then you live in Toledo" is not necessarily true. There is more than one city in Ohio. Choice (B) is incorrect because its converse "If your pet has a tail, it is a cat" is not necessarily true. The pet could be a dog or even some exotic animal such as an ocelot. Choice (C) is incorrect because its converse "If the square of a number is positive, then the number is negative" is not necessarily true. The original number could be positive, since its square would also be positive. Recall that the square of any non-zero number must be positive.

PRACTICE TEST 1

CLEP College Mathematics

Also available at the REA Study Center (*www.rea.com/studycenter*)

This practice test is also offered online at the REA Study Center. All CLEP exams are computer-based, and our test is formatted to simulate test-day conditions. We recommend that you take the online version of the test to receive these added benefits:

- **Timed testing conditions** – helps you gauge how much time you can spend on each question
- **Automatic scoring** – find out how you did on the test, instantly
- **On-screen detailed explanations of answers** – gives you the correct answer and explains why the other answer choices are wrong
- **Diagnostic score reports** – pinpoint where you're strongest and where you need to focus your study

PRACTICE TEST 1

CLEP College Mathematics

(Answer sheets appear in the back of the book.)

TIME: 90 Minutes
60 Questions

Directions: An online scientific calculator will be available for the questions in this test.

Some questions will require you to select from among four choices. For these questions, select the BEST of the choices given.

Some questions will require you to type a numerical answer in the box provided.

Notes: (1) Unless otherwise specified, the domain of any function f is assumed to be the set of all real numbers x for which $f(x)$ is a real number.

(2) i will be used to denote $\sqrt{-1}$

(3) Figures that accompany questions are intended to provide information useful in answering the questions. All figures lie in a plane unless otherwise indicated. The figures are drawn as accurately as possible EXCEPT when it is stated in a specific question that the figure is not drawn to scale.

1. Which one of the following is equivalent to the negation of the statement "Cats are friendly and Bob has a hamster"?

   (A) If cats are friendly, then Bob does not have a hamster.
   (B) If Bob has a hamster, then cats are friendly.
   (C) If cats are not friendly, then Bob has a hamster.
   (D) If Bob does not have a hamster, then cats are not friendly.

2. If A is the set of odd integers, B is the set of multiples of 5, and C is the set of counting numbers, which of the following contains -7?

   (A) $A \cap B$
   (B) C
   (C) A
   (D) $B \cap C$

3. If x is an odd integer and y is even, then which of the following must be an even integer?

   I. $2x + 3y$
   II. xy
   III. $x + y - 1$

   (A) I only
   (B) II only
   (C) I, II, and III
   (D) II and III only

4. An ordinary six-sided cube, with its sides numbered 1 through 6, is rolled twice. The probability of rolling any of the six numbers is equally likely. What is the probability that on two consecutive rolls of the cube, a number less than 3 appears on the first roll and the number 5 appears on the second roll?

5. Not counting the empty set, how many proper subsets are there for R = $\{2, 3, 4\}$?

(A) 5
(B) 6
(C) 7
(D) 8

6. If $f(x) = 2x + 4$ and $g(x) = x^2 - 2$, then $(f \circ g)(x)$, where $(f \circ g)(x)$ is a composition of functions, is

(A) $2x^2 - 8$.
(B) $2x^2 + 8$.
(C) $2x^2$.
(D) $2x^3 + 4x^2 - 4x - 8$.

7. Let P, Q, and R represent statements where P is true, Q is false, and R is false. Which one of the following is a true statement?

(A) (P and R) or Q
(B) (P implies Q) and Not R
(C) Not P or (Q and R)
(D) Not P implies (Q and R)

8. Given that $i = \sqrt{-1}$, what is the simplified expression for $3i^3 - 4i^2 + 5i$?

(A) $-2i - 4$
(B) $-2i + 4$
(C) $2i - 4$
(D) $2i + 4$

9. Consider the function $P(x) = \sqrt{1 - x^2}$ shown below. Which graph represents $P^{-1}\{-1 \leq x \leq 0\}$?

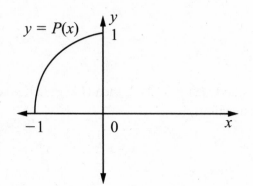

(A)

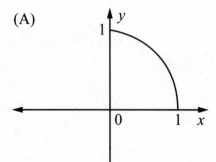

(B)

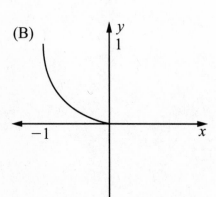

(C)

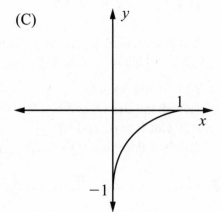

(D)

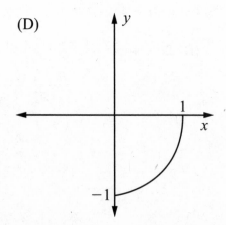

10. What is the range, R, of the function $f(x) = \dfrac{x}{|x|}$?

    (A) $R = \{1, 0\}$
    (B) $R = \{-1, 1\}$
    (C) $(1, -1, 0)$
    (D) All positive integers

11. The sum of three different prime numbers is 22. One of these numbers is 2. What is the largest possible value of either of the other numbers?

12. There are six Knights of the Round Table. Given that Sir Lancelot must sit in a specific chair and that Sir Gawain must be directly on either side of him, in how many ways may the knights be seated?

    (A) 24
    (B) 120
    (C) 48
    (D) 25

13. If $m^x \cdot m^7 = m^{28}$ and $(m^5)^y = m^{15}$, what is the value of $x + y$?

    (A) 31
    (B) 24
    (C) 14
    (D) 12

14. Which one of the following is a factor of $2x^2 - x - 3$?

    (A) $2x - 3$
    (B) $2x + 1$
    (C) $x - 1$
    (D) $x + 3$

15. Suppose $S = \{5, 6, 9\}$ and $T = \{7, 8, 9\}$. Which one of the following ordered pairs is *NOT* in the Cartesian Product of $T \times S$?

    (A) $(9, 9)$
    (B) $(8, 5)$
    (C) $(6, 8)$
    (D) $(7, 6)$

16. If $f(x) = \{(2, 5), (6, 9), (11, 2), (x, 4)\}$ is a function, what is the smallest value x may NOT have?

17. A fair coin is tossed five times. What is the probability of two heads occurring?

    (A) $\dfrac{1}{16}$

    (B) $\dfrac{5}{16}$

    (C) $\dfrac{5}{32}$

    (D) $\dfrac{1}{4}$

18. Given the following list of six numbers:

    $$\pi, \sqrt{5}, \sqrt{\dfrac{4}{25}}, -.212, 5\dfrac{2}{7}, \text{ and } .1\overline{8}$$

    How many of these numbers are irrational?

19. The mean of Sheila's five exam scores is 78. She will be taking three more exams. Assuming that each exam is given the same weight, what must her mean score be on the remaining exams in order to attain a mean score of 84 on all 8 exams?

(A) 94
(B) 92
(C) 90
(D) 88

20. Look at the triangle below.

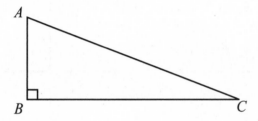

$\overline{AB}$ is perpendicular to $\overline{BC}$. $AB = 10$ and $AC = 26$. What is the area of the triangle?

(A) 260
(B) 240
(C) 130
(D) 120

21. Which graph does NOT represent a function ($y = f(x)$)?

(A)

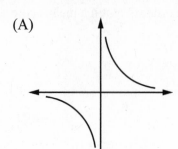

(C)

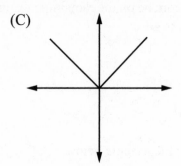

(B)

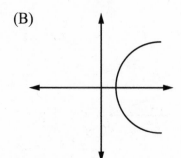

(D)

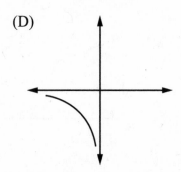

22. Find the minimum value of the function $f(x) = (x - 1)^2 + 3$.

    (A) 0
    (B) 1
    (C) 2
    (D) 3

23. Which one of the following has the lowest value?

    (A) $|-8| - |3|$
    (B) $-|-8 - 3|$
    (C) $|-3 + 8| - |-8 + 3|$
    (D) $-|3 - 8|$

24. Let $U = \{$cat, dog, frog, goat, horse, pig, tiger$\}$, $A = \{$dog, frog, horse, pig$\}$, and $B = \{$dog, goat, pig, tiger$\}$. Define A' as the elements in set U that are not in set A. Which of the following completely describes $A' \cap B$?

   (A) $\{$cat, goat, tiger$\}$
   (B) $\{$goat, tiger$\}$
   (C) $\{$dog, pig$\}$
   (D) $\{$dog, goat, pig$\}$

25. When a positive integer n is divided by 5, the remainder is 4. Which one of the following will yield a remainder of 2 when it is divided by 5?

   (A) $n + 1$
   (B) $n + 2$
   (C) $n + 3$
   (D) $n + 4$

26. Ten white balls and 19 red balls are in a box. If a ball is drawn from the box at random, what are the odds in favor of drawing a red ball?

   (A) $10 : 29$
   (B) $19 : 29$
   (C) $19 : 10$
   (D) $\dfrac{10}{19} : \dfrac{19}{29}$

27. If $A \subset C$ and $B \subset C$, which of the following statements is true?

   (A) The set $A \cup B$ is also a subset of C.
   (B) The complement of A is also a subset of C.
   (C) The complement of B is also subset of C.
   (D) The union of $\overline{A}$ and $\overline{B}$ is also a subset of C.

28. Which one of the following is equivalent to the statement "If Joan sings, then I will play my guitar"?

   (A) Joan sings and I will play my guitar.
   (B) Joan does not sing and I will not play my guitar.
   (C) Joan sings or I will not play my guitar.
   (D) Joan does not sing or I will play my guitar.

29. The function $f(x)$ is divisible by $x + 5$ with no remainder. When $f(x)$ is divided by $x - 1$, the remainder is 3. Which one of the following statements is completely correct?

    (A) 5 is a solution to $f(x) = 0$ and $f(-1) = 3$.
    (B) -5 is a solution to $f(x) = 0$ and $f(1) = 3$.
    (C) 5 is a solution to $f(x) = 0$ and $f(1) = 3$.
    (D) -5 is a solution to $f(x) = 0$ and $f(-1) = 3$.

30. A floor that measures 10 feet by 20 feet is to be tiled with square tiles that are 36 square inches in area. How many tiles are needed to cover the entire floor?

    ┌─────────────────────────────┐
    │   │
    └─────────────────────────────┘

31. How many four-digit numbers are there such that the first digit is odd, the second is even, and there is no repetition of digits?

    (A) 1,200
    (B) 1,625
    (C) 200
    (D) 1,400

32. Which one of the following is the inverse of the statement

    "If animals could talk, then they would not reveal secrets"?

    (A) If animals could talk, then they would reveal secrets.
    (B) If animals could not talk, then they would reveal secrets.
    (C) If animals would not reveal secrets, then they could talk.
    (D) If animals would reveal secrets, then they could not talk.

33. What is the domain of the function given by $f(x) = \dfrac{x^2 + 1}{x + 3}$?

    (A) All numbers except -3
    (B) All numbers except -1
    (C) All numbers except 1
    (D) All numbers except 3

34. Look at the following graph.

Which one of the following represents this graph?

(A) $x \le 2$ or $x > -1$
(B) $-1 < x \le 2$
(C) $x < -1$ or $x \ge 2$
(D) $-1 \le x < 2$

35. What is the median of the following data?

2, 24, 7, 10, 15, 8

(A) 7.5
(B) 8.5
(C) 9
(D) 11

36. Which one of the following is a valid argument?

(A) All rainy days are cloudy.
    Yesterday was not cloudy.
    Yesterday was not rainy.
(B) All trees have brown leaves.
    This plant has brown leaves.
    This plant is a tree.
(C) Some wolves are vicious.
    This animal is vicious.
    This animal is a wolf.
(D) Some people have stocks and bonds.
    Charles has stocks.
    Charles has bonds.

37. The inverse of the function $y = \log_2 \dfrac{2x - 1}{2}$ is

    (A) $y = \dfrac{4^x + 1}{2}$.

    (B) $y = \dfrac{2^{x+1} + 1}{2}$.

    (C) $y = \dfrac{2^{x+1}}{2}$.

    (D) $y = 2^x$.

38. Which one of the following groups of data has exactly two modes?

    (A) 1, 1, 3, 4, 4, 5, 5, 5
    (B) 1, 1, 1, 2, 2, 2, 2
    (C) 1, 2, 3, 3, 4, 4, 5, 5
    (D) 1, 3, 3, 3, 4, 4, 4

39. If f is defined by $f(x) = \dfrac{5x - 8}{2}$ for each real number x, find the solution set for $f(x) > 2x$.

    (A) $\{x \mid x > 6\}$
    (B) $\{x \mid x > 8\}$
    (C) $\{x \mid x < 8\}$
    (D) $\{x \mid 6 < x < 8\}$

40. Look at the following Venn Diagram, for which a Roman numeral has been assigned to each region.

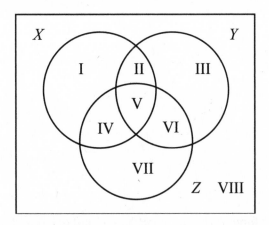

Which region(s) would include $(X \cup Y)'$?

(A) VII and VIII
(B) Only VII
(C) V, VII, and VIII
(D) Only VIII

41. The sample standard deviation, s, of a group of data is given by the formula $s = \sqrt{\dfrac{\sum\limits_{i=1}^{n}(X_i - \overline{X})^2}{n - 1}}$, where X_i represents each data, $\overline{X}$ represents the mean, and n represents the number of data. What is the sample standard deviation for the following data? 4, 5, 6, 9, 11. (Round off your answer to the nearest hundredth).

42. A jar consists of ten marbles, of which seven are green, one is blue, and two are red. Two marbles will be randomly drawn from this jar, one at a time, without replacement. What is the probability of drawing a green marble followed by a red marble?

(A) $\dfrac{83}{90}$

(B) $\dfrac{19}{90}$

(C) $\dfrac{7}{45}$

(D) $\dfrac{7}{50}$

43. If g is a linear function such that $g(4) = 6$ and $g(10) = 21$, what is the value of $g(8)$?

(A) 19
(B) 18
(C) 16
(D) 10

44. Suppose that point K is reflected across the line $y = x$. If the original coordinates of K are $(-9, 3)$, what will be K's new coordinates?

(A) $(-9, -3)$
(B) $(-3, 9)$
(C) $(3, -9)$
(D) $(9, -3)$

45. A group of children share a package of cookies, each having six. If two more children join the group, they can each have four cookies. How many cookies were in the package, assuming none are left over?

(A) 8
(B) 10
(C) 16
(D) 24

46. If x and y are each odd numbers, which one of the following is also an odd number?

   (A) $x + y$
   (B) $4x - 6y$
   (C) $(x)(y) + 10$
   (D) $(x)(y) - 5$

47. A bike wheel has a radius of 12 inches. How many revolutions will it take to cover one mile? (Use 1 mile = 5,280 feet, and $\pi = \dfrac{22}{7}$.)

   (A) 70
   (B) 84
   (C) 120
   (D) 840

48. For which one of the following groups of data are the mean and median identical?

   (A) 2, 2, 5, 6, 8, 8, 11
   (B) 2, 3, 6, 6, 6, 8, 12
   (C) 2, 4, 5, 6, 8, 11
   (D) 2, 5, 5, 7, 8, 12

49. Which one of the following is an equation of a line containing the point $(2, -1)$ and is perpendicular to the graph of $x + 3y = 4$?

   (A) $x + 3y = -1$
   (B) $3x - y = 7$
   (C) $3x + y = 5$
   (D) $x - 3y = 5$

50. A quiz consists of five questions. Three of the questions are true-false, and the other two questions are each multiple-choice with four answer choices. In how many different ways can a student fill in the answers to these questions?

   (A) 14
   (B) 36
   (C) 96
   (D) 128

51. Including the number itself, which one of the following numbers has a total of 72 factors?

(A) $2^2 \times 3^6 \times 5^6$
(B) $2^2 \times 3^2 \times 67^2$
(C) $2^2 \times 5^3 \times 17^1 \times 23^2$
(D) $2^1 \times 5^{19} \times 11^2 \times 43^2$

52. The function $f(x)$ is defined as follows:

$f(x) = 4x - 1$, if $x \le -5$

$\quad\quad = 5x + 1$, if $x > -5$

What is the value of $f(-7) + f(10)$?

53. Given a collection of nine books, in how many different ways can any four of them be placed on a shelf?

(A) 262,144
(B) 60,480
(C) 6561
(D) 3024

54. Which one of the following is true for any function?

(A) A horizontal line may only intersect the graph of the function once.
(B) The inverse must also be a function.
(C) A vertical line may only intersect the graph of the function once.
(D) The function must be defined for all real numbers.

55. Suppose that the point B, which is currently located at $(-4, 5)$, is translated three units to the left and two units up. What is B's new location?

(A) $(-2, 2)$
(B) $(-1, 7)$
(C) $(-7, 7)$
(D) $(-7, 2)$

56. At a social club, there are 20 women and 15 men. Half the women vote Republican and the rest vote Democratic. Only one-fifth of the men vote Republican and the rest vote Democratic. The names of all individuals who vote Republican are placed in a hat. One person's name will be drawn from this hat. What is the probability of drawing the name of a woman?

(A) $\dfrac{10}{13}$

(B) $\dfrac{20}{35}$

(C) $\dfrac{13}{35}$

(D) $\dfrac{10}{35}$

57. A rectangle and a square have the same perimeter. The side of the square is 9 and the length of the rectangle is 13. What is the width of the rectangle?

58. What is the domain of $f(x) = \sqrt{7 - x}$?

(A) All numbers less than or equal to 7
(B) All numbers greater than or equal to 7
(C) All numbers less than or equal to -7
(D) All numbers greater than or equal to -7

59. The number $0.\overline{8}$ is equivalent to what reduced fraction?

60. In a room of 20 people, if each person shakes hands with every other person, how many different handshakes are possible?

(A) 40
(B) 190
(C) 380
(D) 400

PRACTICE TEST 1

Answer Key

1.	(A)	21.	(B)	41.	2.92
2.	(C)	22.	(D)	42.	(C)
3.	(C)	23.	(B)	43.	(C)
4.	$\frac{1}{18}$	24.	(B)	44.	(C)
5.	(B)	25.	(C)	45.	(D)
6.	(C)	26.	(C)	46.	(C)
7.	(D)	27.	(A)	47.	(D)
8.	(D)	28.	(D)	48.	(A)
9.	(D)	29.	(B)	49.	(B)
10.	(B)	30.	800	50.	(D)
11.	17	31.	(D)	51.	(C)
12.	(C)	32.	(B)	52.	22
13.	(B)	33.	(A)	53.	(D)
14.	(A)	34.	(B)	54.	(C)
15.	(C)	35.	(C)	55.	(C)
16.	2	36.	(A)	56.	(A)
17.	(B)	37.	(B)	57.	5
18.	2	38.	(D)	58.	(A)
19.	(A)	39.	(B)	59.	$\frac{8}{9}$
20.	(D)	40.	(A)	60.	(B)

PRACTICE TEST 1

Detailed Explanations of Answers

1. **(A)** When a statement is in the form "If P, then Q," the equivalent state-ment is in the form "Not P or Q." The negation of "Not P or Q" is the statement "P and not Q." Let P represent the statement "Cats are friendly." Let "not Q" represent the statement "Bob has a hamster." Then the given statement in the stem of this question is written in the form "P and not Q," so the negation will be in the form "If P, then Q." Note that Q represents the statement "Bob does not have a hamster."

2. **(C)** To solve this problem we must know the meanings of the rules for the sets that have been stated.

   The set of odd integers is:
   $$A = \{..., -7, -5, -3, -1, 1, 3, 5, 7, ...\}$$
   The set of multiples of 5 is:
   $$B = \{..., -10, -5, 0, 5, 10, ...\}$$
   The set of counting numbers is:
   $$C = \{1, 2, 3, 4, 5, ...\}$$
   We notice that -7 is an element of set A. $A \cap B = \{..., -25, -15, -5, 5, 15, 25, ...\}$, so -7 is not an element of $A \cap B$. Set C does not contain any negative numbers, so -7 is not an element of set C. Finally, $B \cap C = \{5, 10, 15, 20, 25, ...\}$, so -7 is not an element of $B \cap C$.

3. **(C)**

   I.   An odd integer times two will become an even integer. An even integer times any number will remain even. The sum of two even numbers is also an even number.

   Therefore, $2x + 3y$ must be even.

   II.   An even integer times any number will remain even. Therefore, xy must be even.

III.   The sum of an odd integer and an even integer is odd. An odd integer minus one will become even. Therefore, $x + y - 1$ must be even.

4. The correct answer is $\dfrac{1}{18}$. A number less than 3 means 1 or 2. The probability of 1 or 2 appearing on the first roll is $\dfrac{2}{6} = \dfrac{1}{3}$. The probability of 5 appearing on the second roll is $\dfrac{1}{6}$. Since these events are independent, the probability that both will occur is the product of these probabilities, which is $\dfrac{1}{3} \cdot \dfrac{1}{6} = \dfrac{1}{18}$.

5. **(B)**   Given a set of n elements, the number of proper subsets is given by the expression $2^n - 1$. The set of proper subsets does not include the set itself. However the expression $2^n - 1$ does include the empty set. Thus, the answer is $2^n - 2 = 2^3 - 2 = 6$.

6. **(C)** $(f \circ g)(x) = f(x) \circ g(x) = f(g(x))$.

This is the definition of the composition of functions. For the functions given in this problem we have

$$2(g(x)) + 4 = 2(x^2 - 2) + 4$$
$$= 2x^2 - 4 + 4 = 2x^2$$

7. **(D)**   Not P becomes false. Q and R is also false. A statement that reads "False implies false" is always true. Answer choice (A) is wrong because P and R is false, so it reads "False or False." Answer choice (B) is wrong because P implies Q is false, so it reads "False and True," which is false. Answer choice (C) is wrong because Q and R is false, so it reads "False or False."

8. **(D)**

$i^2 = -1$ and $i^3 = i^2(i) = (-1)(i) = -i$. So, $3i^3 - 4i^2 + 5i$
$= (3)(-i) - (4)(-1) + 5i = 5i - 3i + 4 = 2i + 4$.

9. **(D)**   By reflecting $P(x)$ about the line $y = x$, we obtain $P^{-1}(x)$. Graph (D) is the result of such an operation and is therefore the correct choice.

10. **(B)** x can be replaced in the formula $\dfrac{x}{|x|}$ with any real number except 0, so if x is negative, $\dfrac{x}{|x|} = -1$, and if x is positive, $\dfrac{x}{|x|} = 1$. Thus, there are only the two numbers -1 and 1 in the range of our function; $R = \{-1, 1\}$.

11. The correct answer is 17. The sum of the other two numbers must be 20. The possible pairs of integers for which both numbers are prime are 7 and 13 or 3 and 17. Thus, 17 is the largest possible value. Note that 1 and 19 cannot be considered since 1 is not a prime.

12. **(C)** Since Sir Lancelot must sit in an assigned chair, and Sir Gawain on either side of him, there are 4! or 24 ways of seating the other four. For each of these arrangements, Sir Gawain can be in either of two seats, so the total number of ways of seating the knights is 24×2 or 48.

13. **(B)** For the first equation, $x + 7 = 28$, so $x = 21$. For the second equation, $5y = 15$, so $y = 3$. Then $x + y = 24$.

14. **(A)** In factored form, $2x^2 - x - 3 = (2x - 3)(x + 1)$. The other correct factor would be $x + 1$.

15. **(C)** The Cartesian Product of $T \times S$ consists of all ordered pairs, where the first element is chosen from T and the second element is chosen from S. The correct answer choice $(6, 8)$ is not a member of $T \times S$. It is a member of $S \times T$.

16. The answer is 2. If $f(x)$ is a function, then each member of the domain is mapped into one and only one member of the range. Hence, x may not be 2, 6, or 11; these numbers are already first components of ordered pairs in $f(x)$.

17. **(B)** With each toss there are two possibilities, heads or tails. There are five tosses so there are $2^5 = 32$ possible outcomes. The number of ways to choose two heads from five tosses is

$$\binom{5}{2} = \frac{5!}{2!3!} = 10.$$

So the probability is

$$\frac{10}{32} = \frac{5}{16}.$$

18. The correct answer is 2. The irrational numbers are π and $\sqrt{5}$. Irrational numbers cannot be written as a quotient of two integers. The other four numbers can be written as a quotient of integers.

$$\sqrt{\frac{4}{25}} = \frac{2}{5}, \quad -.212 = -\frac{212}{1000}, \quad 5\frac{2}{7} = \frac{37}{7}, \quad \text{and} \quad .1\overline{8} = \frac{17}{90}.$$

19. **(A)** The total number of points on Sheila's five exams is $(5)(78) = 390$. In order to attain a mean score of 84 on all the exams, she will need a total of $(8)(84) = 672$ points. Thus, she needs a total of $672 - 390 = 282$ points on the next three exams. Finally $282 \div 3 = 94$.

20. **(D)** Using the Pythagorean Theorem, $AB^2 + BC^2 = AC^2$. By substitution, $10^2 + BC^2 = 26^2$. Then $BC^2 = 676 - 100 = 576$, so $BC = 24$. The area of the triangle is given by the formula $\left(\frac{1}{2}\right)(AB)(BC) = \left(\frac{1}{2}\right)(10)(24) = 120$.

21. **(B)** To determine whether or not a graph represents a function, it is possible to apply the "vertical line test." If any vertical line to the graph intersects it in more than one point, the graph is not a function. The only graph for which a vertical line passes through more than one point is (B). A relation is a function if for any x there is one and only one y.

22. **(D)** Since $(x - 1)^2 > 0$ for all x, the minimum value occurs when $(x - 1)^2 = 0$. Hence, $f(x) = 0 + 3 = 3$ is the minimum value. This conclusion can also be made by considering the graph of the function, which is a parabola.

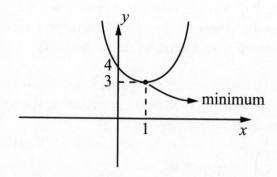

23. **(B)** $-|-8 - 3| = -|-11| = -11$. The values of answer choices (A), (C), and (D) are 5, 0, and -5, respectively.

24. **(B)** In this example, $A' = \{$cat, goat, tiger$\}$.

Then, $A' \cap B = \{$cat, goat, tiger$\} \cap \{$dog, goat, pig, tiger$\} = \{$goat, tiger$\}$.

25. **(C)** From the first sentence, we can deduce that when $n + 1$ is divided by 5, there is no remainder. To find an expression that would have a remainder of 2 when it is divided by 5, simply add 2 to $n + 1$. Then, $n + 1 + 2 = n + 3$.

26. **(C)** If an event can happen in p ways and fail to happen in q ways, then, if $p > q$, the odds are p to q in favor of the event happening.

If $p < q$, then the odds are q to p against the event happening. In this case, $p = 19$ and $q = 10$, $p > q$. Thus, the odds in favor of the event of drawing a red ball are $19 : 10$.

27. **(A)** The set $A \cup B$ contains all elements which belong to either set A or set B. Since all elements, which belong to set A or set B, also belong to C, the set $A \cup B$ is a subset of C.

28. **(D)** Given a conditional statement in the form "If P then Q," an equivalent statement is in the form "Not P or Q." In this example, P is the statement "Joan sings" and Q is the statement "I will play my guitar."

29. **(B)** By the Factor Theorem, whenever $x - c$ is a factor of $f(x)$, the number c must be a solution to $f(x) = 0$. In this case, $c = -5$. Thus, -5 must be a solution to $f(x) = 0$. Also, by the Remainder Theorem, whenever a function $f(x)$ is divided by $x - c$, the remainder must be $f(c)$. In this case, $c = 1$, so that the remainder, which is 3, must be the value of $f(1)$.

30. The answer is 800. The floor that measures 10 ft. by 20 ft. has an area of $10 \times 20 = 200$ sq. ft. The tiles with 36 sq. in. of area must measure 6 in. by 6 in. or $\frac{1}{2}$ ft. by $\frac{1}{2}$ ft. for $\frac{1}{4}$ sq. ft. of area. Because it would take four tiles to cover 1 sq. ft., $4 \times (200 \text{ sq. ft.}) = 800$ tiles would be needed to cover the entire floor.

31. **(D)** There are five possibilities for the first (1, 3, 5, 7, 9), five for the second (0, 2, 4, 6, 8), eight possibilities for the third $(10 - 2)$, and seven for the fourth $(8 - 1) = 7$. We subtracted 2 possibilities on the third and 3 on the fourth because there is no repetition, as shown below:

$$5 \times 5 \times (10 - 2)(8 - 1) = 5 \times 5 \times 8 \times 7 = 1,400$$

32. **(B)** Given a statement in the form "If P, then Q," the inverse is in the form "If not P, then not Q." In this example P is represented by "animals could talk" and Q is represented by "they would not reveal secrets."

33. **(A)** The domain refers to all allowable values of x. In this example, the only value(s) of x that are not allowed are those for which the denominator is zero. If $x + 3 = 0$, then $x = -3$. Thus, -3 is the only value that is not allowed in the domain.

34. **(B)** The shaded area lies between -1 and 2. Since there is a dot at 2, this number is included. Likewise, since there is an open circle at -1, this number is not included.

35. **(C)** To find the median, first arrange the data in ascending order. The data will then appear as follows: 2, 7, 8, 10, 15, 24. The median will be the average of the two middle numbers. Thus, the median is $\dfrac{(8 + 10)}{2} = 9$.

36. **(A)** An argument is valid if given that the premises are true, then the conclusion <u>must</u> be true. Only answer choice (A) would satisfy this definition. Answer choice (B) is wrong because other objects besides trees may have brown leaves. Answer choice (C) is wrong because animals other than wolves may be vicious. Answer choice (D) is wrong because people may own only stocks, only bonds, both stocks and bonds, or neither stocks nor bonds.

37. **(B)** To determine the inverse of a function, it is necessary to replace the variable y by x and x by y.

The function $y = \log_2 \dfrac{2x - 1}{2}$ becomes:

$$x = \log_2 \frac{2y - 1}{2}$$

$$2^x = \frac{2y - 1}{2}$$

$$2^x \cdot 2 = 2y - 1$$

$$2^{x+1} = 2y - 1$$

$$2^{x+1} + 1 = 2y$$

$$y = \frac{2^{x+1} + 1}{2}$$

38. **(D)** The correct answer (D) has exactly two modes, namely 3 and 4. Answer choice (A) has a single mode of 5. Answer choice (B) has a single mode of 2. Answer choice (C) has three modes, namely 3, 4, and 5.

39. **(B)** To find the solution set of $f(x) > 2x$, we proceed as follows:

$$\frac{5x - 8}{2} > 2x$$

$$5x - 8 > 4x$$

$$-8 > -x$$

which implies $x > 8$.

40. **(A)** $(X \cup Y)'$ means the regions that are <u>not</u> included by X, by Y, or by both X and Y. Note that answer choice (C) is wrong because it includes region V, which is in all three of X, Y, and Z.

41. The correct answer is 2.92. The value of $\overline{X}$ is $(4 + 5 + 6 + 9 + 11) \div 5 = 7$.

The value of $\sum_{i=1}^{n} (X_i - \overline{X})^2$ can be found by computing

$$(4 - 7)^2 + (5 - 7)^2 + (6 - 7)^2 + (9 - 7)^2 + (11 - 7)^2$$
$$= 9 + 4 + 1 + 4 + 16 = 34.$$

Then $s = \sqrt{\dfrac{34}{4}} = \sqrt{8.5} \approx 2.92$

42. **(C)** The probability of drawing a green marble is $\dfrac{7}{10}$. Since the second marble is drawn from among the remaining nine marbles, the probability of drawing a red marble is $\dfrac{2}{9}$. The probability of both occurrences is $\left(\dfrac{7}{10}\right)\left(\dfrac{2}{9}\right) = \dfrac{7}{45}$.

43. **(C)** Since g is a linear function, the ratio of the change in $g(x)$ values to the change in x values between any two points on the graph is constant. For the two given points $(4, 6)$ and $(10, 21)$, the constant ratio is $(21 - 6) \div (10 - 4) = 2.5$. Let $g(8) = k$. Combining this point with $(4, 6)$, we can state that $(k - 6) \div (8 - 4) = 2.5$. Simplifying, we get $\dfrac{k - 6}{4} = 2.5$. Then, multiplying both sides of the equation by 2.5, $k - 6 = 10$, so $k = 16$.

44. **(C)** When any point (x, y) is reflected across the line $y = x$, the coordinates are simply reversed so that the new coordinates become (y, x). Thus, $(-9, 3)$ becomes $(3, -9)$.

45. **(D)** Using some algebraic expressions to represent the unknown should help here. Let:

$$x = \text{the number of children in the original group}$$
$$6x = \text{the number of cookies in the package.}$$

When two more children join the group, the expressions are:

$$x + 2 = \text{the number of children}$$
$$4(x + 2) = \text{the number of cookies in the package.}$$

Notice that there are two expressions for "the number of cookies in the package." These are equivalent expressions, so set up an equation and solve for x.

$$6x = 4(x + 2)$$
$$6x = 4x + 8$$
$$2x = 8$$
$$x = 4$$

Then $6x = 24$, which is the number of cookies.

46. **(C)** The product of two odd numbers is an odd number. When an odd number is added to 10, the result is still an odd number. Each of answer choices (A), (B), and (D) results in an even number.

47. **(D)** The circumference of the wheel is:

$$C = 2\pi(1 \text{ ft.})$$
$$C = 2\left(\frac{22}{7}\right) = \frac{44}{7} \text{ ft.}$$

To find the number of revolutions the wheel takes, calculate:

$$5,280 \div \frac{44}{7} = 5,280 \times \frac{7}{44}$$
$$= 120 \times 7 = 840 \text{ revolutions}$$

48. **(A)** The mean and the median are each 6. For answer choice (B), is 6 but the mean is $\dfrac{43}{7}$. For answer choice (C), the mean is 6 but the is 5.5. For answer choice (D), the median is 6 but the mean is 6.5.

49. **(B)** Rewriting $x + 3y = 4$ as $y = -\dfrac{1}{3}x + \dfrac{4}{3}$, we can identify the slope as $-\dfrac{1}{3}$. A line that is perpendicular to the graph of this line must have a slope that is the negative reciprocal of $-\dfrac{1}{3}$, which is 3. When answer choice (B) is rewritten as $y = 3x - 7$, the slope can be identified as 3. Note also that $y = 3x - 7$ contains the point $(2, -1)$. The slopes for answer choices (A), (C), and (D) are $-\dfrac{1}{3}$, -3, and $\dfrac{1}{3}$ respectively.

50. **(D)** The number of different ways to answer the true-false questions is $2^3 = 8$. The number of different ways to answer the multiple-choice questions is $4^2 = 16$. Then the number of different ways to fill in all five answers is $(8)(16) = 128$.

51. **(C)** The total number of factors can be found as follows: After the given number is written in prime factorization form, add 1 to each exponent. Then take the product of these numbers. For answer choice (C), the computation would be $(2 + 1)(3 + 1)(1 + 1)(2 + 1) = (3)(4)(2)(3) = 72$. The total number of factors for answer choices (A), (B), and (D) are 147, 27, and 360 respectively.

52. The correct answer is 22. $f(-7) = (4)(-7) - 1 = -29$ and $f(10) = (5)(10) + 1 = 51$. Then $-29 + 51 = 22$.

53. **(D)** There are 9 selections for the first spot on the shelf, 8 selections for the second spot, 7 selections for the third spot, and 6 selections for the fourth spot. The number of different ways is $(9)(8)(7)(6) = 3024$. This is a permutation of 9 items taken 4 at a time.

54. **(C)** Since each x value in the domain can correspond only to one y value in the range, any vertical line can intersect the graph at most once. Answer choice (A) is wrong because the same y value may correspond to two

different x values. An example to show why answer choice (B) is wrong would be $f(x) = \{(1, 2), (2, 3), (4, 3)\}$. $f(x)$ is a function consisting of three points. Its inverse would be $\{(2, 1), (3, 2), (3, 4)\}$, which is not a function. An example to show why answer choice (D) is wrong would be $g(x) = \{(1, 5), (3, 7)\}$. $g(x)$ is only defined for $x = 1$ or $x = 3$.

55. **(C)** We need to subtract 3 from the first coordinate and add 2 to the second coordinate. The new location for point B is $(-4-3, 5+2) = (-7, 7)$.

56. **(A)** There are $\left(\dfrac{1}{2}\right)(20) = 10$ women who vote Republican, and there are $\left(\dfrac{1}{5}\right)(15) = 3$ men who vote Republican. Thus, there are a total of 13 people who vote Republican and whose names are in the hat. From these 13 names, 10 are women. Thus, the required probability is $\dfrac{10}{13}$.

57. The correct answer is 5. The perimeter of the square is $(4)(9) = 36$, which is the same as the perimeter of the rectangle. The perimeter of a rectangle is twice the length plus twice the width. Twice the length is 26, so twice the width must be 10. Thus, the width is 5.

58. **(A)** The domain is defined as all numbers for which $7 - x$ is at least zero. Solving $7 - x \geq 0$, we get $x \leq 7$.

59. The correct answer is $\dfrac{8}{9}$. Let $N = 0.\overline{8}$. Multiply both sides of the equation by 10 to get $10N = 8.\overline{8}$. Subtract $N = 0.\overline{8}$ from $10N = 8.\overline{8}$ to get $9N = 8$.

Then $N = \dfrac{8}{9}$.

60. **(B)** Each handshake involves two people, so the number of handshakes for 20 people is given by the expression $_{20}C_2 = (20)(19) \div 2 = 190$.

PRACTICE TEST 2

CLEP College Mathematics

Also available at the REA Study Center (*www.rea.com/studycenter*)

This practice test is also offered online at the REA Study Center. All CLEP exams are computer-based, and our test is formatted to simulate test-day conditions. We recommend that you take the online version of the test to receive these added benefits:

- **Timed testing conditions** – helps you gauge how much time you can spend on each question
- **Automatic scoring** – find out how you did on the test, instantly
- **On-screen detailed explanations of answers** – gives you the correct answer and explains why the other answer choices are wrong
- **Diagnostic score reports** – pinpoint where you're strongest and where you need to focus your study

PRACTICE TEST 2

CLEP College Mathematics

(Answer sheets appear in the back of the book.)

TIME: 90 Minutes
60 Questions

Directions: An online scientific calculator will be available for the questions in this test.

Some questions will require you to select from among four choices. For these questions, select the BEST of the choices given.

Some questions will require you to type a numerical answer in the box provided.

Notes: (1) Unless otherwise specified, the domain of any function f is assumed to be the set of all real numbers x for which $f(x)$ is a real number.

(2) i will be used to denote $\sqrt{-1}$

(3) Figures that accompany questions are intended to provide information useful in answering the questions. All figures lie in a plane unless otherwise indicated. The figures are drawn as accurately as possible EXCEPT when it is stated in a specific question that the figure is not drawn to scale.

1. What is the domain of the function defined by $f(x) = \sqrt{-x + 1} + 5$?

   (A) $\{x|x \geq 0\}$
   (B) $\{x|x \leq 1\}$
   (C) $\{x|0 \ x \leq x \leq 1\}$
   (D) $\{x|x \geq -1\}$

2. If a fair coin is tossed three times, what is the probability of getting a head on the first toss, a tail on the second toss, and a head on the third toss?

    (A) $\dfrac{1}{64}$

    (B) $\dfrac{1}{16}$

    (C) $\dfrac{1}{8}$

    (D) $\dfrac{1}{4}$

3. A counting number with exactly two different factors is called a prime number. Which of the following pairs of numbers are consecutive prime numbers?

    (A) 27 and 29
    (B) 31 and 33
    (C) 41 and 43
    (D) 37 and 39

4. Which one of the following is equivalent to the statement "If roses are red, then the sky is blue"?

    (A) Roses are red and the sky is blue.
    (B) Roses are not red and the sky is blue.
    (C) Roses are red or the sky is blue.
    (D) Roses are not red or the sky is blue.

5. Local officials were concerned about the speeds of cars that passed a certain intersection in the small town of Speedville, so they monitored the speed of a randomly chosen group of 50 cars at this intersection. Following are the results.

SPEED OF CAR	NUMBER OF CARS
31 – 35 miles per hour	6
36 – 40 miles per hour	7
41 – 45 miles per hour	10
46 – 50 miles per hour	16
51 – 55 miles per hour	11

Based on the table shown above, what percent of the cars exceeded 40 miles per hour?

6. How many proper subsets are there for $\{1, 3, 5, 7, ..., 19\}$?

(A) 361
(B) 1023
(C) 30,321
(D) 524,287

7. If a and b are odd integers, which of the following must be an even integer?

I. $\dfrac{a + b}{2}$

II. $ab - 1$

III. $\dfrac{ab + 1}{2}$

(A) I only
(B) II only
(C) I and II only
(D) II and III only

8. Which graph below represents $x < -1$ or $x \geq 3$?

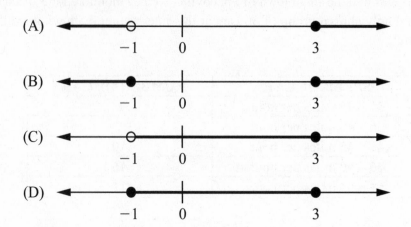

9. What is the contrapositive of the statement "If today is Tuesday, then we are going to Chicago"?

(A) If today is not Tuesday, then we are not going to Chicago.
(B) If we are going to Chicago, then today is Tuesday.
(C) If today is Tuesday, then we are not going to Chicago.
(D) If we are not going to Chicago, then today is not Tuesday.

10. If $\log_{10} x = 3$ and $\log_{10} .01 = y$, what is the value of $x + y$?

11. Which one of the following numbers is divisible by both 3 and 4, but is *NOT* divisible by 8?

(A) 24
(B) 36
(C) 48
(D) 56

12. Five hundred people were surveyed concerning their favorite snack. The pie chart shown below is a summary of the data collected.

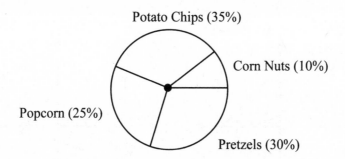

Potato Chips (35%)

Corn Nuts (10%)

Popcorn (25%)

Pretzels (30%)

What is the combined total of people who chose either pretzels or potato chips as their favorite snack?

(A) 425
(B) 325
(C) 300
(D) 250

13. The area of a trapezoid is given by the formula $A = \left(\dfrac{1}{2}\right)(h)(b_1 + b_2)$, where h represents the height, b_1 represents one base, and b_2 represents the other base. What is the area of a trapezoid in which the height is 5 and the mean value of the two bases is 7?

14. Which one of the graphs below represents the inverse of the function $y = 3x + 4$?

(A)

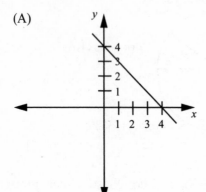

(C)

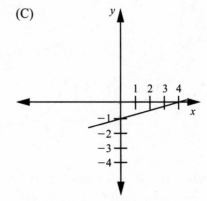

(B)

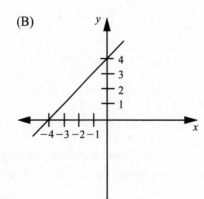

(D)

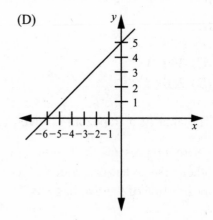

15. Given sets T, V such that $T \cap V = \{3, 7\}$ and $T \cup V = \{1, 3, 5, 7, 9, 11\}$, which one of the following could *NOT* represent T?

(A) $\{1, 3\}$
(B) $\{3, 7\}$
(C) $\{1, 3, 7\}$
(D) $\{3, 5, 7\}$

16. On which interval is the function below positive?

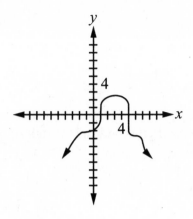

(A) [0, 2]
(B) (0, 2)
(C) (1, 5)
(D) [1, 5]

17. Which one of the following is the conjugate for the complex number $-5 + 8i$?

(A) $5 - 8i$
(B) $-5 - 8i$
(C) $5 + 8i$
(D) $-8 + 5i$

18. A telephone number consists of seven digits including zero. How many different telephone numbers exist if each digit appears only one time in the number?

(A) $7!$

(B) $\dfrac{10!}{3!}$

(C) 70

(D) 7^7

19. If $f(x) = 2x^2 + 5$ and $g(x) = -x - 6$, what is the value of $f(g(-1))$?

   (A) -93
   (B) -45
   (C) 13
   (D) 55

20. Using only the digits 1, 2, 3, 4, 5, how many different four-digit numbers are possible if the left-most digit is odd and the right-most digit is even? Repetition of digits is allowed.

   (A) 150
   (B) 100
   (C) 48
   (D) 36

21. Which one of the following numbers has exactly 4 prime factors?

   (A) 1155
   (B) 945
   (C) 625
   (D) 375

22. Twelve more than twice a number is 31 less than three times the number. Find the number.

   (A) -43
   (B) 43
   (C) -9
   (D) 19

23. Let P, Q, and R each represent false statements. Which one of the following statements is also false?

   (A) P implies (Q and R)
   (B) (P and Q) or (not R)
   (C) P or (Q implies not R)
   (D) P and (not Q or R)

24. What is the value of $|-(-3) - (15)| + |-6^2 + 6|$?

```

```

25. For which one of the following functions is the inverse *NOT* a function?

    (A) $\{(1, 2), (3, 4), (7, 6)\}$
    (B) $\{(1, 1), (3, 3), (5, 5)\}$
    (C) $\{(4, 2), (2, 4), (6, 6)\}$
    (D) $\{(3, 5), (4, 4), (6, 5)\}$

26. How many composite numbers are there between 30 and 40 that also end in an odd digit?

    (A) 2
    (B) 3
    (C) 4
    (D) 5

27. What is the least common multiple of $18x$ and $24xy$?

    (A) 2
    (B) $72xy$
    (C) $6x$
    (D) 48

28. Let P and Q represent statements. In which one of the following is P a necessary condition for Q?

    (A) P implies Q
    (B) P and Q
    (C) Q implies P
    (D) Q or P

29. If $Pr(A) = 0.2$ (the probability of event A is $\frac{1}{5}$), $Pr(B) = 0.6$, and $Pr(A \cap B) = 0.1$, what is $Pr(A \cup B)$?

```

```

30. Look at the following Venn Diagram, where J and K represent sets.

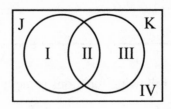

Which region(s) represent K'?

(A) I and IV
(B) Only IV
(C) Only I
(D) III and IV

31. If the graph of a function $f(x)$ crosses the x-axis twice, which one of the following statements is true?

(A) There are exactly two distinct real values for $f(x) = 0$.
(B) There is exactly one real value for $f(x) = 0$.
(C) There are at most two distinct real values for $f(x) = 0$.
(D) There are at least two distinct real values for $f(x) = 0$.

32. If $h(x) = -5x^2 - x + 9$, what is the value of $h\left(\dfrac{1}{2}\right)$?

(A) 3.25
(B) 5.25
(C) 7.25
(D) 9.25

33. Triangles ABC and $A'B'C'$ are similar right triangles. Find the length of side AC.

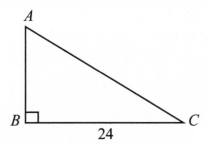 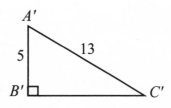

(A) 10
(B) 12
(C) 13
(D) 26

34. Professor Leonardo gave a quiz to the 20 students in her class. After calculating the mean grade to be 80, she realized that she mistakenly wrote 95 for one of her students. If this student's test score was really 55, what was the actual mean grade for the class?

(A) 78
(B) 77
(C) 76
(D) 75

35. There are 4 men and 6 women in a room. Two people will be randomly selected. What is the probability that both will be men?

(A) $\dfrac{2}{3}$

(B) $\dfrac{2}{5}$

(C) $\dfrac{2}{11}$

(D) $\dfrac{2}{15}$

36. Let $S = \{x \mid x$ is an integer greater than 2$\}$, let $T = \{x \mid x$ is an odd negative integer$\}$, and let $V = \{x \mid x$ is a non-zero integer with an absolute value less than 4$\}$. Which one of the following is equivalent to the empty set?

    (A) $T \cap V$
    (B) $S \cap V$
    (C) $S \cap T$
    (D) $T \cup V$

37. Which one of the following is an irrational number?

    (A) $\sqrt{.0025}$

    (B) $.\overline{124}$

    (C) $.26226222622226....$

    (D) $\dfrac{11}{7}$

38. For a group of data, arranged in ascending order, the median is located midway between the twelfth and thirteenth numbers. How many data are in this group?

    ┌─────────────────────────┐
    │   │
    └─────────────────────────┘

39. Which one of the following functions has a range of all numbers except zero?

    (A) $f(x) = x - 1$

    (B) $f(x) = \dfrac{1}{x}$

    (C) $f(x) = \dfrac{1}{\sqrt{x}}$

    (D) $f(x) = -|x|$

40. Using an ordinary deck of 52 playing cards, what is the probability of drawing three black cards in a row, without replacement?

    (A) $\dfrac{1}{8}$

    (B) $\dfrac{2}{17}$

    (C) $\dfrac{3}{25}$

    (D) $\dfrac{4}{33}$

41. Let $F = \{3, 6, 9\}$, $G = \{4, 6, 8\}$, and $H = \{6, 8, 9\}$. The ordered pair $(3, 6)$ belongs to which of the following Cartesian products?

    (A) $F \times G$ and $F \times H$
    (B) Only $F \times G$
    (C) Only $F \times H$
    (D) $F \times H$ and $G \times H$

42. If m and n are consecutive integers, and $m < n$, which one of the following statements is always true?

    (A) $n - m$ is even.
    (B) m must be odd.
    (C) $m^2 + n^2$ is even.
    (D) $n^2 - m^2$ is odd.

43. The points $(-2, 6)$ and $(1, 11)$ lie on the graph of a linear function. Which one of the following points must also lie on the graph of this function?

    (A) $(2, 14)$
    (B) $(3, 15)$
    (C) $(4, 16)$
    (D) $(5, 17)$

44. Which one of the following is the negation for the statement "Some women enjoy shopping"?

    (A) Some women do not enjoy shopping.
    (B) At least one woman enjoys shopping.
    (C) No women enjoy shopping.
    (D) All women enjoy shopping.

45. The graph of two linear equations in x and y contains two perpendicular lines. Which system below could represent the equations of these two lines?

    (A) $4x + 7y = 11$
        $8x + 14y = 13$

    (B) $5x + 11y = 17$
        $11x + 5y = 20$

    (C) $3x + 7y = 10$
        $14x - 6y = 15$

    (D) $9x + 2y = 19$
        $18x - 4y = 21$

46. What is the smallest positive number such that when it is divided by 9 or by 12, the remainder is 1?

47. Which one of the following functions is identical to its inverse?

    (A) $y = x^2$
    (B) $x + y = 10$
    (C) $2x + y = 12$
    (D) $y = x^2 + x$

48. Let J represent any non-empty set. Consider the following group of combinations of sets:

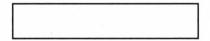

$$J \cup \varnothing, \; J \cap \varnothing, \; J - J', \; J \cap J', \; J - \varnothing, \; \varnothing - J$$

How many of these six combinations are equivalent to $\varnothing$?

49. For which one of the following data groups is the mean and median identical, but there is <u>no</u> mode?

(A) 2, 2, 4, 4, 4, 6, 6
(B) 2, 4, 6, 8, 10, 12
(C) 2, 5, 7, 9, 10, 13
(D) 2, 3, 5, 5, 6, 7, 7

50. Suppose the function $f(x)$ contains the points $(-1, 3)$, $(0, 4)$ and $(2, 7)$. If a new function $g(x)$ is created by moving each point of $f(x)$ two units to the left and three units down, which one of the following points <u>must</u> $g(x)$ contain?

(A) $(-1, 0)$
(B) $(1, 4)$
(C) $(0, 10)$
(D) $(-2, 1)$

51. On a bookshelf, there are 25 books, 10 of which are red and the remaining books are green. Twelve of these books are math books of which 4 are red. How many green non-math books are there?

(A) 7
(B) 8
(C) 11
(D) 15

52. Four less than three times x is greater than 6. Find all values of x.

    (A) $x < \dfrac{10}{3}$

    (B) $x > \dfrac{10}{3}$

    (C) $x < 5$

    (D) $x > \dfrac{2}{3}$

53. Which one of the following represents a biconditional statement for P and Q?

    (A) (P or Q) and (P or not Q)
    (B) (P implies Q) or (Q implies P)
    (C) (P and Q) or (Q and not P)
    (D) (P implies Q) and (Q implies P)

54. The number n is known to have exactly two composite factors, not counting itself. Which one of the following could be the value of n?

    (A) 30
    (B) 20
    (C) 15
    (D) 8

55. In how many different ways can all the letters of the word LEASES be arranged without repeating the same sequence of letters?

    (A) 180
    (B) 360
    (C) 540
    (D) 720

56. The formula for the area of a circle is $A = \pi r^2$, where r represents the radius. How is the area of a given circle affected if the radius is doubled?

    (A) The area is doubled.
    (B) The area is tripled.
    (C) The area is quadrupled.
    (D) The change in area cannot be determined without knowing the value of the radius.

57. Suppose the probability that it will rain today is .8 and the probability that Ken will go bowling today is .6. Assuming that these events are independent, what is the probability that Ken will *NOT* go bowling today and it will also *NOT* rain today?

(A) .70
(B) .40
(C) .20
(D) .08

58. Which one of the following functions shows that both the domain and the range are all non-zero numbers?

(A) $f(x) = \sqrt{x}$

(B) $f(x) = \dfrac{2}{x}$

(C) $f(x) = |x|$

(D) $f(x) = x - 3$

59. An equilateral triangle and a rectangle have identical perimeters. Each side of the triangle is 13 and the width of the rectangle is 8. What is the rectangle's length?

60. Suppose a number m is divisible by 6 and by 8. Which one of the following statements is true?

(A) m must be divisible by any multiple of 48.
(B) m must be divisible by any multiple of 16.
(C) m must be divisible by 3 and by 4.
(D) m must be divisible by any prime number.

PRACTICE TEST 2

Answer Key

1.	(B)	21.	(A)	41.	(A)
2.	(C)	22.	(B)	42.	(D)
3.	(C)	23.	(D)	43.	(C)
4.	(D)	24.	42	44.	(C)
5.	74	25.	(D)	45.	(C)
6.	(B)	26.	(B)	46.	37
7.	(B)	27.	(B)	47.	(B)
8.	(A)	28.	(C)	48.	3
9.	(D)	29.	0.7	49.	(B)
10.	998	30.	(A)	50.	(D)
11.	(B)	31.	(A)	51.	(A)
12.	(B)	32.	(C)	52.	(B)
13.	35	33.	(D)	53.	(D)
14.	(C)	34.	(A)	54.	(B)
15.	(A)	35.	(D)	55.	(A)
16.	(C)	36.	(C)	56.	(C)
17.	(B)	37.	(C)	57.	(D)
18.	(B)	38.	24	58.	(B)
19.	(D)	39.	(B)	59.	11.5
20.	(A)	40.	(B)	60.	(C)

PRACTICE TEST 2

Detailed Explanations of Answers

1. **(B)** The only restriction for the domain is that $-x + 1$ must be greater than or equal to zero.

   $-x + 1 \geq 0$
   $\Rightarrow 1 \geq x$, which means $x \leq 1$.

2. **(C)** The probability of getting a head = Pr(head) $= \dfrac{1}{2}$

   The probability of getting a tail = Pr(tail) $= \dfrac{1}{2}$

   Pr(head, tail, head) = Pr(head on 1st) $\times$ Pr(tail on 2nd) $\times$ Pr(head on 3rd)

   $$= \left(\frac{1}{2}\right) \times \left(\frac{1}{2}\right) \times \left(\frac{1}{2}\right) = \frac{1}{8}$$

3. **(C)** To test whether a number, N, is prime, we need to test if N is divisible by any of the prime numbers $\{2, 3, 5, 7, 11, 13, \ldots\}$ up to the largest natural number, k, whose square is less than or equal to the number we are testing, N. If N is divisible by any of the prime numbers $P \leq k$, where $k^2 \leq N$, then N is not a prime number. If N is not divisible by any of the prime numbers $P \leq k$, where $k^2 \leq N$, then N is a prime number. For example, to test whether 29 is a prime number, we need to test if 29 is divisible by any of the prime numbers starting with 2 and up to 5, that is, we test if 29 is divisible by 2, 3, or 5, since $6^2 = 36$ is > 29. Since 29 is not divisible by any of these primes, 29 is a prime number.

   Since all the prime numbers, except 2, are odd, it follows that the difference between any two consecutive prime numbers is 2. Thus, each of the pairs of numbers given in the answer choices as possible answers are two consecutive odd numbers.

   Thus, to answer this question, we need to test if any of the pairs of numbers given in the answer choices is a pair of prime numbers. Testing these pairs of numbers yields:

(A) 27 and 29. Since $6^2 = 36$, and since $36 > 27$, and $36 > 29$, we need to test if 27 or 29 is divisible by any of the prime numbers less than or equal to 5. That is, if 27 or 29 is divisible by 2, 3, or 5. Since 27 is divisible by 3, then 27 and 29 is not a pair of consecutive prime numbers.

(B) 31 and 33. Again, $6^2 = 36$, $36 > 31$, and $36 > 33$. Hence, we need to test if 31 or 33 is divisible by any of the prime numbers less than or equal to 5. That is, if 31 or 33 is divisible by any of the primes 2, 3, or 5. Since 33 is divisible by 3, then 31 and 33 is not a pair of two consecutive prime numbers.

(C) 41 and 43. Since $7^2 = 49$, $49 > 41$, and $49 > 43$, it follows that we need to test if 41 or 49 is divisible by any of the prime numbers less than or equal to 6. That is, if 41 or 49 is divisible by any of the primes 2, 3, or 5. Since 41 is not divisible by any of these three primes and 43 is not divisible by any of these three primes either, it follows that 41 and 43 is a pair of consecutive prime numbers.

(D) 37 and 39. Since 39 is divisible by 3, then 37 and 39 is not a pair of prime numbers.

4. **(D)** When a statement is in the form "If P, then Q," the equivalent statement is in the form "Not P or Q." Let P represent the statement "Roses are red" and let Q represent the statement "The sky is blue." Then "Not P" is represented by "Roses are not red."

5. The correct answer is 74. The number of cars that were traveling in excess of 40 miles per hour can be tabulated using the figures in the lowest three rows. So, $10 + 16 + 11 = 37$ cars were traveling faster than 40 miles per hour. Then, $\left(\dfrac{37}{50}\right)(100)\,\% = 74\%$.

6. **(B)** By simply counting, we can determine that there are 10 elements in this set. The number of proper subsets is given by the expression $2^n - 1$, where n represents the number of elements. Thus, the number of proper subsets is $2^{10} - 1 = 1023$.

7. **(B)** We can express odd integers as $a = 2x + 1$ and $b = 2y + 1$ where x and y are integers.

I. $\dfrac{a + b}{2} = \dfrac{(2x + 1) + (2y + 1)}{2}$

$= \dfrac{2x + 2y + 2}{2}$

$= x + y + 1$,

which is not necessarily even.

II. $ab - 1 = (2x + 1)(2y + 1) - 1$

$= (4xy + 2x + 2y + 1) - 1$

$= 2(2xy + x + y)$,

which is always even (divisible by 2).

III. $\dfrac{ab + 1}{2} = \dfrac{(4xy + 2x + 2y + 1) + 1}{2}$

$= 2xy + x + y + 1$,

which is not necessarily even.

8. **(A)** The expression $x < -1$ means all numbers less than but not including -1, which is represented by the ray on the left. The expression $x \geq 3$ means all numbers greater than and including 3, which is represented by the ray on the right.

9. **(D)** Given a statement in the form "If P then Q," the contrapositive is the statement "If not Q, then not P." In this example, P represents "Today is Tuesday" and Q represents "we are going to Chicago."

10. The correct answer is 998. $\log_{10} x = 3$ means $x = 10^3 = 1000$ and $\log_{10} .01 = y$ means $10^y = .01 = 10^{-2}$, so $y = -2$. Then $x + y = 1000 - 2 = 998$.

11. **(B)** The number 36 is divisible by both 3 and 4, since it is a multiple of 12. However, 36 is not divisible by 8.

12. **(B)** The total percent of people who chose either pretzels or potato chips as their favorite snack is 30% + 35% = 65%. This percent corresponds to (.65)(500) = 325 people.

13. The correct answer is 35. The easiest way to calculate this answer is to rewrite the area formula as $A = \left(\dfrac{1}{2}\right)(b_1 + b_2)(h)$. The expression $\left(\dfrac{1}{2}\right)(b_1 + b_2)$ represents the mean of b_1 and b_2. Then the area is $(7)(5) = 35$.

14. **(C)** The inverse of the function $f(x) = 3x + 4$ is obtained by replacing y by x and x by y.

    $y = 3x + 4$
    becomes
    $x = 3y + 4 \Rightarrow 3y = x - 4 \Rightarrow y = \dfrac{x - 4}{3}$

    The graph of $y = \dfrac{x - 4}{3}$ is given below:

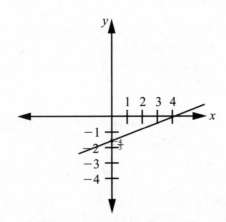

15. **(A)** Since $T \cap V = \{3, 7\}$, T <u>must</u> contain at least the elements 3 and 7. Answer choice (A) does not contain the element 7.

16. **(C)** A function is positive when the points on its graph lie above the x-axis. In our graph this occurs when x is between 1 and 5 or (1, 5).

17. **(B)** For any complex number in the form $a + bi$, the conjugate is given by the expression $a - bi$. In this example, $a = -5$ and $b = 8$.

18. **(B)** There are ten integers from 0 to 9; hence, there are ten choices for the first digit of the phone number. Since no digit can be repeated, we have only nine choices for the second digit. Similarly, there are eight choices for the third digit and so on. Since there are seven digits in a phone number, we have

$$10 \times 9 \times 8 \times 7 \times 6 \times 5 \times 4 = \frac{10!}{3!} \text{ possible phone numbers.}$$

19. **(D)**

$$g(-1) = -(-1) - 6 = 1 - 6 = -5.$$

Then, $f(-5) = (2)(-5)^2 + 5 = (2)(25) + 5 = 55.$

20. **(A)** The number of choices for the left-most digit is 3, since there are three odd digits (1, 3, 5). The number of choices for the right-most digit is 2, since there are two even digits (2, 4). For each of the other two digits there are 5 choices, since repetition of any of the five digits is allowed. Then $(3)(5)(5)(2) = 150$ represents the number of possibilities when creating a four-digit number.

21. **(A)** In prime factorization form, $1155 = (3)(5)(7)(11), 945 = (3)^3(5)(7)$, $625 = (5)^4$, and $375 = (3)(5)^3$. Of these, only the number 1155 has 4 prime factors.

22. **(B)** Let x represent the number. The corresponding equation to the statement "Twelve more than twice a number is 31 less than three times the number" is $2x + 12 = 3x - 31$. By subtracting $3x$ from each side, the equation becomes $-x + 12 = -31$. Now subtract 12 from each side. The equation simplifies to $-x = -43$. Since $-x$ means $-1x$, the last step is to divide each side by -1.

Then $x = \dfrac{-43}{-1} = 43.$

23. **(D)** Since P is false, a compound statement with false and any truth value must be false. The statement "Not Q or R" is actually true, but "false and true" is equivalent to false. In answer choice (A), it would read: "false implies false," which is true. In answer choice (B), it would read: "false

or true," which is true. In answer choice (C), it would read: "false or (false implies true)," which becomes "false or true," which is true. Recall that "false implies any truth value" is always true.

24. The correct answer is 42. $| - (-3) - (15) | = | 3 - 15 | = | - 12 | = 12$ and $| - 6^2 + 6 | = | - 36 + 6 | = | - 30 | = 30$. Then $12 + 30 = 42$.

25. **(D)** The inverse of any function is found by reversing the domain and range. For answer choice (D), the inverse would be $\{(5, 3), (4, 4), (5, 6)\}$. This is not a function because the number 5 in the domain is paired with both 3 and 6 in the range. For the inverse of each of the other answer choices, each number in the domain will be paired with only one number in the range.

26. **(B)** The three composite numbers between 30 and 40 that end in an odd digit are 33, 35, and 39. Note that the numbers 31 and 37 are prime numbers.

27. **(B)** The least common multiple or LCM of $18x$ and $24xy$ is the smallest number that both of these numbers can divide into evenly without a remainder. To find the LCM, prime factorize $18x$ and $24xy$ into:

$$18x = (2)(3)(3)(x)$$
$$24xy = (2)(2)(2)(3)(x)(y)$$

Now, take the largest group of each number and variable represented. Of the 2's, the larger group is $(2)(2)(2)$ from the $24xy$. Of the 3's, the larger group is the $(3)(3)$ from the $18x$. There is also one x and y to be represented. Therefore, the LCM of $18x$ and $24xy$ is $(2)(2)(2)(3)(3)(xy) = 72xy$.

The LCM may also be calculated by a modified form of long division, as follows:

$$
\begin{array}{c|cc}
6x & 18x & 24xy \\
\hline
3x & 3 & 4y \\
\hline
4y & 1 & 4y \\
\hline
& 1
\end{array}
$$

Then $(6x)(3)(4y) = 72xy$.

28. **(C)** "P is a necessary condition for Q" is equivalent to "If Q, then P," which means "Q implies P."

29. The answer is 0.7. The probability of the union of two events equals the sum of the probabilities of each event separately minus the probability of the intersection.

$$\Pr(A \cup B) = \Pr(A) + \Pr(B) - \Pr(A \cap B)$$
$$= 0.2 + 0.6 - 0.1 = 0.7$$

30. **(A)** K' means the region(s) outside the circle shown as K, which would be both I and IV.

31. **(A)** The number of distinct real values for $f(x) = 0$ can be found by looking at the number of times that the graph of $f(x)$ crosses the x-axis.

32. **(C)**

$$h\left(\frac{1}{2}\right) = (-5)\left(\frac{1}{2}\right)^2 - \frac{1}{2} + 9 = -\frac{5}{4} - \frac{1}{2} + 9 = 7\frac{1}{4} = 7.25$$

33. **(D)** The sides of similar triangles are proportional. Using the Pythagorean Theorem, the length of $B'C'$ is:

$$5^2 + (B'C')^2 = 13^2$$
$$25 + (B'C')^2 = 169$$
$$(B'C')^2 = 144$$
$$B'C' = 12$$

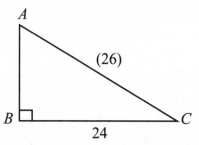
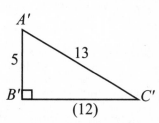

BC and $B'C'$ are corresponding sides of similar triangles in a ratio of 24:12 or 2:1. Similarly, side AC and $A'C'$ have the same ratio; AC must be twice as long as $A'C'$. In other words, $AC = 2(13) = 26$.

34. **(A)** The original total points for the class was $(80)(20) = 1600$. Since one of the students was erroneously given 40 extra points, the total points for the class should have been 1560. The correct mean for the class is $1560 \div 20 = 78$.

35. **(D)** The number of combinations of selecting two men is $_4C_2 = \dfrac{(4)(3)}{2!}$ $= 6$ The number of combinations of selecting any two people is $_{10}C_2 = \dfrac{(10)(9)}{2!} = 45$. Then the probability is given by $\dfrac{6}{45} = \dfrac{2}{15}$.

36. **(C)** $S = \{3, 4, 5, 6, \ldots\}$, $T = \{-1, -3, -5, -7, \ldots\}$, and $V = \{-1, -2, -3, 1, 2, 3\}$. We note that S and T have no common elements, so their intersection is the empty set. $T \cap V = \{-1, -3\}$ and $S \cap V = \{3\}$.

37. **(C)** An irrational number is one that cannot be written in the form $\dfrac{P}{Q}$, where P and Q are integers. Any decimal that does not end or repeat is always an irrational number. Answer choice (C) shows a definite pattern, but the number of 2's between the 6's never repeats. Answer choice (A) can be written as .05 or $\dfrac{1}{20}$. Answer choice (B) can be written as $\dfrac{124}{999}$, and answer choice (D) is already written in the form $\dfrac{P}{Q}$, where $P = 11$ and $Q = 7$.

38. The correct answer is 24. For a group of n data, arranged in ascending order, the median is located in the $\left(\dfrac{n+1}{2}\right)^{th}$ position. In this example, the median is located in the 12.5^{th} position. Then $\dfrac{n+1}{2} = 12.5$, which leads to $n + 1 = 25$. So, $n = 24$.

39. **(B)** The fraction $\dfrac{1}{x}$ is defined except when $x = 0$. Since 1 divided by a non-zero number can be any non-zero value, the range of this function is all numbers except zero. For answer choice (A), the range is all numbers. For answer choice (C), the range is all positive numbers. For answer choice (D), the range is all non-positive numbers.

40. **(B)**

$$\text{Pr(black card on first draw)} = \frac{26}{52}$$

$$\text{Pr(black card on second draw)} = \frac{25}{51}$$

$$\text{Pr(black card on third draw)} = \frac{24}{50}$$

$$\text{Pr(drawing three straight black cards)} = \frac{26}{52} \times \frac{25}{51} \times \frac{24}{50} = \frac{2}{17}$$

41. **(A)** The Cartesian product of any two sets is formed using ordered pairs in which the first element comes from the first set and the second element comes from the second set. Since the number 3 comes from set F and the number 6 comes from each of sets G and H, the ordered pair (3, 6) belongs to both $F \times G$ and $F \times H$.

42. **(D)** If m and n are consecutive integers, and $m < n$, it follows that

$$n = m + 1$$

Now, we can check each of the answer choices (A) through (D) as follows:

(A) $n - m = (m + 1) - m = m + 1 - m = 1$,
which is odd. Thus, the statement in answer choice (A) is false.
(B) Since no specific information is given about the integer m, m can be an odd integer or an even integer. So, the statement in answer choice (B) is false.

$$\text{(C) } m^2 + n^2 = m^2 + (m + 1)^2 = m^2 + m^2 + 2m + 1$$
$$= 2m^2 + 2m + 1$$
$$= 2(m^2 + m) + 1$$

Since 2 times any integer (even or odd) yields an even integer, it follows that $2(m^2 + m)$ is an even integer, and hence $2(m^2 + m) + 1$ is an odd integer. Hence, the statement in answer choice (C) is false.

$$\text{(D) } n^2 - m^2 = (m + 1)^2 - m^2 = m^2 + 2m + 1 - m^2$$
$$= 2m + 1$$

Again, since 2 times any integer (even or odd) yields an even integer, it follows that $2m$ is an even integer and $2m + 1$ is always an odd integer. Hence, the statement in answer choice (D) is correct.

43. **(C)** Since the graph represents a linear function, the slope must be constant between any two points. Based on the two given points, the slope is

$\dfrac{11-6}{1-(-2)}=\dfrac{5}{3}$. Using the point (4, 16) in conjunction with either (−2, 6) or (1, 11), the slope would also be $\dfrac{5}{3}$. As an example, using (4, 16) and (−2, 6), the slope is $\dfrac{6-16}{-2-4}=\dfrac{-10}{-6}=\dfrac{5}{3}$. If any of the points in answer choices (A), (B), or (D) were paired with either (−2, 6) or with (1, 11), the slope would not be $\dfrac{5}{3}$.

44. **(C)** The statement "Some women enjoy shopping" is equivalent to the statement "At least one woman enjoys shopping." The negation of this statement would mean that it is not true that at least one woman enjoys shopping, which becomes the statement "No women enjoy shopping."

45. **(C)** If two lines are perpendicular, their slopes must be negative reciprocals of each other. For the equation $3x+7y=10$, it can be rewritten as $y=-\dfrac{3}{7}x+\dfrac{10}{7}$, so its slope is $-\dfrac{3}{7}$. For the equation $14x-6y=15$, it can be rewritten as $y=\dfrac{14}{6}x-\dfrac{15}{6}=\dfrac{7}{3}x-\dfrac{5}{2}$, so its slope is $\dfrac{7}{3}$. Since $\dfrac{7}{3}$ is the negative reciprocal of $-\dfrac{3}{7}$, these two lines are perpendicular. In answer choice (A), the lines are parallel, since their slopes are each $-\dfrac{4}{7}$. For answer choice (B), the slopes are $-\dfrac{5}{11}$ and $-\dfrac{11}{5}$, so the lines are neither parallel nor perpendicular. For answer choice (D), the slopes are $-\dfrac{9}{2}$ and $\dfrac{9}{2}$, so the lines are neither parallel nor perpendicular.

46. The correct answer is 37. First find the least common multiple of 9 and 12. Rewrite each number in prime factorization form. $9=3^2$ and $12=2^2\bullet3$. The least common multiple is $2^2\bullet3^2=36$. This means that 36, when divided by 9 or by 12, leaves no remainder. It follows that when 37 is divided by 9 or by 12, the remainder is 1.

47. **(B)** To find the inverse of any function, interchange the variables. In answer choice (B), the equation remains the same. In each of answer choices (A), (C), and (D), the equation will not be identical with the interchange of x and y.

48. The correct answer is 3, namely $J \cap \varnothing$, $J \cap J'$, and $\varnothing - J$. Note that $J \cup \varnothing$ is equivalent to J, $J - J'$ is equivalent to J, and $J - \varnothing$ is equivalent to J.

49. **(B)** The mean is $\dfrac{2 + 4 + 6 + 8 + 10 + 12}{6} = \dfrac{42}{6} = 7$ and the median is $\dfrac{6 + 8}{2} = 7$. Furthermore, there is no mode, since each number appears only once. Answer choice (A) is wrong because even though both the mean and median are 4, the number 4 is also the mode. Answer choice (C) is wrong because the mean is $\dfrac{46}{6}$, whereas the median is 8. Answer choice (D) is wrong because even though both the mean and median are 5, both the numbers 5 and 7 are modes.

50. **(D)** Since $(0, 4)$ is on the graph of $f(x)$, the graph of $g(x)$ must contain $(0 - 2, 4 - 3) = (-2, 1)$. Based on the method of creating $g(x)$, the only other two points that $g(x)$ must contain would be $(-3, 0)$, and $(0, 4)$.

51. **(A)** There are $25 - 10 = 15$ green books. The number of green math books is $12 - 4 = 8$. Then the number of green non-math books is $15 - 8 = 7$.

52. **(B)** Set up the inequality in mathematical symbols and solve.

$$3x - 4 > 6$$
$$3x > 6 + 4$$
$$3x > 10$$
$$x > \dfrac{10}{3}$$

53. **(D)** This is the definition of a biconditional statement involving P and Q.

54. **(B)** The factors of 20 are 1, 2, 4, 5, 10, and 20. Not counting the number itself, the two composite factors are 4 and 10. Answer choice (A) is wrong because 30 has composite factors of 6, 10, and 15. Answer choice (C) is wrong because 15 has no other composite factors besides itself. Answer choice (D) is wrong because 8 has only one other composite factor besides itself, namely 4.

55. **(A)** The number of different arrangements of these letters is given by the expression $\dfrac{6!}{2! \cdot 2!} = \dfrac{720}{4} = 180$. Recall that the $n! = (n)(n-1)(n-2)(...)(1)$, where n is any whole number.

56. **(C)** When the radius is doubled, the area is multiplied by 4. To check this, suppose the radius were 3. The area is $(\pi)(3^2) = 9\pi$. If the radius doubles to 6, the area becomes $(\pi)(6^2) = 36\pi$, which is four times as large.

57. **(D)** The probability that it will not rain today is $1 - .8 = .2$, and the probability that Ken will not go bowling today is $1 - .6 = .4$. Since these events are independent, the probability that both will occur is given by $(.2)(.4) = .08$.

58. **(B)** For $f(x) = \dfrac{2}{x}$, the domain is only restricted by the fact that x cannot equal zero. The value of $\dfrac{2}{x}$ is any number except zero, since the quotient of 2 and a non-zero number must be a non-zero number. Answer choice (A) is wrong because the domain and range are non-negative numbers. Answer choice (C) is wrong because the range is non-negative numbers, but the domain is all real numbers. Answer choice (D) is wrong because the domain and range are all real numbers.

59. The correct answer is 11.5. The perimeter of the equilateral triangle is $(13)(3) = 39$, which is also the perimeter of the rectangle. The formula for the perimeter of a rectangle is $P = 2L + 2W$, where L is the length and W is the width. We know that the width is 8, so substitution into the perimeter formula gives $39 = 2L + (2)(8)$. This leads to $2L = 39 - 16 = 23$. Thus, $L = 11.5$.

60. **(C)** Since m is divisible by 6, it must also be divisible by any factor of 6. Since m is divisible by 8, it must also be divisible by any factor of 8. We know that 3 is a factor of 6 and 4 is a factor of 8, so answer choice (C) is correct. The number 72 would show why answer choices (A), (B), and (D) are wrong. For answer choice (D), note that 72 is not divisible by a prime such as 5, for example.

ANSWER SHEETS

Practice Test 1
Practice Test 2

PRACTICE TEST 1

Answer Sheet

1. Ⓐ Ⓑ Ⓒ Ⓓ
2. Ⓐ Ⓑ Ⓒ Ⓓ
3. Ⓐ Ⓑ Ⓒ Ⓓ
4. [   ]
5. Ⓐ Ⓑ Ⓒ Ⓓ
6. Ⓐ Ⓑ Ⓒ Ⓓ
7. Ⓐ Ⓑ Ⓒ Ⓓ
8. Ⓐ Ⓑ Ⓒ Ⓓ
9. Ⓐ Ⓑ Ⓒ Ⓓ
10. Ⓐ Ⓑ Ⓒ Ⓓ
11. [   ]
12. Ⓐ Ⓑ Ⓒ Ⓓ
13. Ⓐ Ⓑ Ⓒ Ⓓ
14. Ⓐ Ⓑ Ⓒ Ⓓ
15. Ⓐ Ⓑ Ⓒ Ⓓ
16. [   ]
17. Ⓐ Ⓑ Ⓒ Ⓓ
18. [   ]
19. Ⓐ Ⓑ Ⓒ Ⓓ
20. Ⓐ Ⓑ Ⓒ Ⓓ

21. Ⓐ Ⓑ Ⓒ Ⓓ
22. Ⓐ Ⓑ Ⓒ Ⓓ
23. Ⓐ Ⓑ Ⓒ Ⓓ
24. Ⓐ Ⓑ Ⓒ Ⓓ
25. Ⓐ Ⓑ Ⓒ Ⓓ
26. Ⓐ Ⓑ Ⓒ Ⓓ
27. Ⓐ Ⓑ Ⓒ Ⓓ
28. Ⓐ Ⓑ Ⓒ Ⓓ
29. Ⓐ Ⓑ Ⓒ Ⓓ
30. [   ]
31. Ⓐ Ⓑ Ⓒ Ⓓ
32. Ⓐ Ⓑ Ⓒ Ⓓ
33. Ⓐ Ⓑ Ⓒ Ⓓ
34. Ⓐ Ⓑ Ⓒ Ⓓ
35. Ⓐ Ⓑ Ⓒ Ⓓ
36. Ⓐ Ⓑ Ⓒ Ⓓ
37. Ⓐ Ⓑ Ⓒ Ⓓ
38. Ⓐ Ⓑ Ⓒ Ⓓ
39. Ⓐ Ⓑ Ⓒ Ⓓ
40. Ⓐ Ⓑ Ⓒ Ⓓ

41. [   ]
42. Ⓐ Ⓑ Ⓒ Ⓓ
43. Ⓐ Ⓑ Ⓒ Ⓓ
44. Ⓐ Ⓑ Ⓒ Ⓓ
45. Ⓐ Ⓑ Ⓒ Ⓓ
46. Ⓐ Ⓑ Ⓒ Ⓓ
47. Ⓐ Ⓑ Ⓒ Ⓓ
48. Ⓐ Ⓑ Ⓒ Ⓓ
49. Ⓐ Ⓑ Ⓒ Ⓓ
50. Ⓐ Ⓑ Ⓒ Ⓓ
51. Ⓐ Ⓑ Ⓒ Ⓓ
52. [   ]
53. Ⓐ Ⓑ Ⓒ Ⓓ
54. Ⓐ Ⓑ Ⓒ Ⓓ
55. Ⓐ Ⓑ Ⓒ Ⓓ
56. Ⓐ Ⓑ Ⓒ Ⓓ
57. [   ]
58. Ⓐ Ⓑ Ⓒ Ⓓ
59. [   ]
60. Ⓐ Ⓑ Ⓒ Ⓓ

PRACTICE TEST 2

Answer Sheet

1. Ⓐ Ⓑ Ⓒ Ⓓ
2. Ⓐ Ⓑ Ⓒ Ⓓ
3. Ⓐ Ⓑ Ⓒ Ⓓ
4. Ⓐ Ⓑ Ⓒ Ⓓ
5. [＿＿＿＿＿]
6. Ⓐ Ⓑ Ⓒ Ⓓ
7. Ⓐ Ⓑ Ⓒ Ⓓ
8. Ⓐ Ⓑ Ⓒ Ⓓ
9. Ⓐ Ⓑ Ⓒ Ⓓ
10. [＿＿＿＿＿]
11. Ⓐ Ⓑ Ⓒ Ⓓ
12. Ⓐ Ⓑ Ⓒ Ⓓ
13. [＿＿＿＿＿]
14. Ⓐ Ⓑ Ⓒ Ⓓ
15. Ⓐ Ⓑ Ⓒ Ⓓ
16. Ⓐ Ⓑ Ⓒ Ⓓ
17. Ⓐ Ⓑ Ⓒ Ⓓ
18. Ⓐ Ⓑ Ⓒ Ⓓ
19. Ⓐ Ⓑ Ⓒ Ⓓ
20. Ⓐ Ⓑ Ⓒ Ⓓ

21. Ⓐ Ⓑ Ⓒ Ⓓ
22. Ⓐ Ⓑ Ⓒ Ⓓ
23. Ⓐ Ⓑ Ⓒ Ⓓ
24. [＿＿＿＿＿]
25. Ⓐ Ⓑ Ⓒ Ⓓ
26. Ⓐ Ⓑ Ⓒ Ⓓ
27. Ⓐ Ⓑ Ⓒ Ⓓ
28. Ⓐ Ⓑ Ⓒ Ⓓ
29. [＿＿＿＿＿]
30. Ⓐ Ⓑ Ⓒ Ⓓ
31. Ⓐ Ⓑ Ⓒ Ⓓ
32. Ⓐ Ⓑ Ⓒ Ⓓ
33. Ⓐ Ⓑ Ⓒ Ⓓ
34. Ⓐ Ⓑ Ⓒ Ⓓ
35. Ⓐ Ⓑ Ⓒ Ⓓ
36. Ⓐ Ⓑ Ⓒ Ⓓ
37. Ⓐ Ⓑ Ⓒ Ⓓ
38. [＿＿＿＿＿]
39. Ⓐ Ⓑ Ⓒ Ⓓ
40. Ⓐ Ⓑ Ⓒ Ⓓ

41. Ⓐ Ⓑ Ⓒ Ⓓ
42. Ⓐ Ⓑ Ⓒ Ⓓ
43. Ⓐ Ⓑ Ⓒ Ⓓ
44. Ⓐ Ⓑ Ⓒ Ⓓ
45. Ⓐ Ⓑ Ⓒ Ⓓ
46. [＿＿＿＿＿]
47. Ⓐ Ⓑ Ⓒ Ⓓ
48. [＿＿＿＿＿]
49. Ⓐ Ⓑ Ⓒ Ⓓ
50. Ⓐ Ⓑ Ⓒ Ⓓ
51. Ⓐ Ⓑ Ⓒ Ⓓ
52. Ⓐ Ⓑ Ⓒ Ⓓ
53. Ⓐ Ⓑ Ⓒ Ⓓ
54. Ⓐ Ⓑ Ⓒ Ⓓ
55. Ⓐ Ⓑ Ⓒ Ⓓ
56. Ⓐ Ⓑ Ⓒ Ⓓ
57. Ⓐ Ⓑ Ⓒ Ⓓ
58. Ⓐ Ⓑ Ⓒ Ⓓ
59. [＿＿＿＿＿]
60. Ⓐ Ⓑ Ⓒ Ⓓ

Glossary

Absolute inequality: An absolute inequality for the set of real numbers means that for any real value for the variable, x, the sentence is always true.

Absolute value: The absolute value of a number is the distance the number is from the zero point on the number line. The absolute value of a number or an expression is always greater than or equal to zero (i.e., nonnegative).

Acute triangle: A triangle whose interior angles are all acute.

Adding: Increasing in amount, number, or degree.

Addition: A mathematical process to combine numbers and/or variables into an equivalent quantity, number, or algebraic expression.

Addition property of inequality: For all numbers a, b, and c, the following are true: (1) If $a > b$, then $a + c > b + c$ and $a - c > b - c$; (2) If $a < b$, then $a + c < b + c$ and $a - c < b - c$. In other words, if the same number or expression is added or subtracted from both sides of a true inequality, the new inequality is also true.

Additive inverse: The opposite of a given number.

Altitude: The height of an object or point in relation to sea level or ground level.

Altitude of the trapezoid: The distance between the bases of a trapezoid.

Angle bisector: The division of something into two equal or congruent parts, usually by a line, which is then called a bisector.

Antecedent: In the compound statement "if a, then b," statement (a) is called the antecedent.

Arc: The portion of a circle cut off by a central angle.

Area of a circle: The area of a circle is found using the formula $A = \pi r^2$.

Area of a square: The area of a square is found using the formula $A = s^2$.

Area of a triangle: The area of a triangle is found using the formula $A = \frac{1}{2}bh$.

Associative property of addition: The sum of any three real numbers is the same, regardless of the way they are grouped.

Associative property of multiplication: The product of any three real numbers is the same, regardless of the way they are grouped.

Bar graphs: A graph that uses horizontal or vertical bars to display countable data.

Base: A number used as a repeated factor.

Base angles: A pair of angles that include only one of the parallel sides.

Base of the triangle: The base of a triangle can be any side of a triangle, but specifically it is the side that is perpendicular to the height.

Bell-shaped graph: A graph that is approximately normal, i.e., mean = median = mode.

Biconditional: The statement of the form "p if and only if q."

Bimodal: When a set of data has two modes.

Bivariate data: Shows the relationship between two variables.

Cartesian product: Given two sets M and N, the Cartesian product, denoted as $M \times N$, is the set of all ordered pairs of elements in which the first component is a member of M and the second component is a member of N.

Center: The middle point of a circle or sphere, equidistant from every point on the circumference or surface.

Central angle: An angle whose vertex is at the center of a circle and whose sides are radii.

Central tendency: Mean, median, and mode, which describe the tendency (the "middle" or the "center") of a set of data.

Chord: A line segment joining two points on a circle.

Circle: The set of all points in a plane that are equidistant from a fixed point called the center.

Circle graphs: A graph that uses sections of a circle to show how portions of a set of data compare with the whole and with other parts of the data.

Circumference: The length of the outer edge of a circle.

Circumscribed circle: A circle passing through all the vertices of a polygon.

Closed interval: A set of real numbers that contains its endpoints.

Combination: An arrangement of items, events, or people from a set, without regard to the order.

Commutative property of addition: The sum of two real numbers is the same even if their positions are changed.

Commutative property of multiplication: The product of two real numbers is the same even if their positions are changed.

Complex number: A number of the form $a + bi$ where a and b are real numbers.

Composite function: Composition of functions is the process of combining two functions where one function is performed first and the result of which is substituted in place of each x in the other function.

Composite numbers: The set of integers, other than 0 and 1, that are not prime.

Concentric circles: Circles that have the same center and unequal radii.

Conclusion: The implication, "then b"; the end or finish of an event or process.

Conditional inequality: An inequality whose validity depends on the values of the variables in the sentence; that is, certain values of the variables will make the sentence true and others will make it false.

Conditional probability: The probability that event B occurs, given that event A has occurred.

Conditional statement: The compound statement "if a, then b."

Congruent circles: Circles whose radii are congruent.

Conjunction: A statement using "and."

Conjunction operator: The symbol for "and" in a conjunction ($\land$).

Consecutive angles: Two angles that have their vertices at the endpoints of the same side of a parallelogram.

Consecutive integers: The set of integers that differ by 1: $\{n, n + 1, n + 2, ...\}$ ($n =$ an integer).

Consequent: In the compound statement "if a, then b," statement b is called the consequent.

Consistent: A system of equations is said to be consistent when it has at least one ordered pair that satisfies both equations.

Contrapositive: The contrapositive of an implication is formed by negating each statement, then swapping the order of the negated statements.

Converse: The converse of an implication is formed by switching the hypothesis and conclusion of a conditional statement.

Counting rule: If one experiment can be performed in m ways, and a second experiment can be performed in n ways, then there are $m \times n$ distinct ways both experiments can be performed in this specified order. The counting principle can be applied to more than two experiments.

Data analysis: The process of evaluating data using analytical and logical reasoning; often involves putting numerical values into picture form, such as bar graphs, line graphs, and circle graphs.

Deduction: The last part of a syllogism is a statement to the effect that the general statement which applies to the group also applies to the individual.

Deductive reasoning: The technique of employing a syllogism to arrive at a conclusion.

De Morgan's laws for sentences: Two laws, one stating that the denial of the conjunction of a class of implications is equivalent to the disjunction of the denials of an implication, and the other stating that the denial of the disjunction of a class of implications is equivalent to the conjunction of the denials of the implications.

Dependent equations: Equations that represent the same line; therefore, every point on the line of a dependent equation represents a solution.

Dependent variable: A variable (often denoted by y) whose value depends on that of another.

Diagonal: A line segment joining any two nonconsecutive sides of a polygon.

Diameter: A chord that passes through the center of the circle.

Difference of two sets: The difference of two sets, A and B, written as $A - B$, is the set of all elements that belong to A but do not belong to B.

Direction: The direction of a scatter plot tells what happens to the response variables as the explanatory variable increases.

Discriminant: The value of $b^2 - 4ac$ in a quadratic equation.

Disjoint: Two sets A and B are disjoint if they have no elements in common (their intersection is the null set).

Disjunction: Two statements a and b, shown by the compound statement "a or b."

Disjunction operator: The symbol for "or" in a conjunction ($\vee$).

Distributive laws: The property that states that multiplying a sum by a number is the same as multiplying each addend by the number and then adding the products. The distributive property says that if a, b, and c are real numbers, then: $a \times (b + c) = (a \times b) + (a \times c)$.

Distributive property: A theorem asserting that one operator can validly be distributed over another.

Domain: The set of all the values of x in a relation.

Element: Each individual item belonging to a set is called an element or member of that set.

Empty set: A set with no members.

Equal sets: Two sets are equal if they have exactly the same elements.

Equation: Two algebraic expressions separated by an equal sign that means that the two sides have equal value.

Equiangular: A triangle with each angle equaling 60 degrees.

Equilateral triangle: A triangle in which all three sides are equal.

Equivalent expressions: Expressions such as $x + 7$ and $7 + x$, which have equal values for any value of x.

Equivalent inequalities: Inequalities that have the same solution set.

Equivalent sets: Two sets are equivalent if they have the same number of elements.

Even integers: The set of integers divisible by 2: $\{\dots, -4, -2, 0, 2, 4, 6, \dots\}$.

Explanatory variable: Used to predict the response variable.

Exponent: Power that indicates the number of times the base appears when multiplied by itself.

Exponential function: Functions that contain a variable in an exponent.

Exterior angle: An angle formed outside a triangle by one side of the triangle and the extension of an adjacent side.

Factor: Any counting number that divides into another number with no remainder is called a factor of that number.

Factor theorem: If $x = c$ is a solution of the equation $f(x) = 0$, then $(x - c)$ is a factor of $f(x)$.

Factorial: The factorial of an integer is the product of that number and all the integers less than it down to 1, or n!

Finite set: A set where the number of its elements can be counted.

Fraction: Rational number which consist of a numerator (on the top) and a denominator (on the bottom).

Function: A relation that assigns each input exactly one output; any process that assigns a single value of y to each number of x.

Fundamental counting principle: If one experiment can be performed in m ways, and a second experiment can be performed in n ways, then there are $m \times n$ distinct ways both experiments can be performed in this specified order.

Fundamental theorem of algebra: Every polynomial equation $f(x) = 0$ of degree greater than zero has at least one root either real or complex.

Graph of an inequality: One variable is represented by either a ray or a line segment on the real number line.

Half-open interval: A continuous set of real numbers that contains only one endpoint.

Height: A line segment from a vertex of the triangle perpendicular to the opposite side.

Histogram: A graph that displays the frequency of a set of data in equal-size intervals, or ranges of data; is an appropriate display for quantitative data; is used primarily for continuous data, but it may be used for discrete data that have a wide spread.

Hypotenuse: The side opposite the right angle in a right triangle.

Hypothesis: In the compound statement "if a, then b," "if a" is called the hypothesis.

Identity property of addition: The sum of zero and any number is that number.

Identity property of multiplication: The product of any number and 1 is that number.

Imaginary number: The square root of -1.

Implication: The compound statement "if a, then b," also called a conditional statement.

Improper fractions: A fraction that has a numerator greater than or equal to its denominator.

Inconsistent: A system of equations with no solution.

Independent: The outcome of one event does not influence the outcome of the other event.

Independent variable: A variable whose value does not depend on the value of another variable.

Inequality: A mathematical statement that two quantities are not or may not be equal.

Infinite set: Any set that is not finite, that is, has a countless number of elements.

Inscribed angle: An angle whose vertex is on the circle and whose sides are chords of the circle.

Integers: The set of numbers represented as $\{..., -3, -2, -1, 0, 1, 2, 3, ...\}$.

Interior angle: An angle formed by two sides of a triangle and includes the third side within its collection of points.

Intersection of two sets: The intersection of two sets A and B, denoted $A \cap B$, is the set of all elements that belong to both A and B.

Intuition: Intuition is the process of making generalizations on insight.

Inverse: A function that has the range of the original function as its domain and the domain of the original function as its range.

Irrational numbers: A number that cannot be expressed as the ratio of two integers.

Isosceles trapezoid: A trapezoid whose nonparallel sides are equal.

Isosceles triangle: A triangle that has at least two equal sides.

Legs: The two sides that are not the hypotenuse of a right triangle.

Line graphs: Data display that shows points connected by line segments; often used to display data that change over time.

Line of centers: The line passing through the centers of two (or more) circles.

Linear equation: An equation with a graph that is a straight line.

Logarithmic function: Equations written in the form $y = \log_b x$ where $b > 0$ and $b \neq 1$.

Logical equivalence: Statements that are either both true or both false.

Logical value: A sentence X is true (or T) if X is true, and false (or F) if X is false.

Logically equivalent: Two sentences are logically equivalent if and only if it is impossible for one of the sentences to be true while the other sentence is false; that is, if and only if it is impossible for the two sentences to have different truth values.

Logically false: A sentence is logically false if and only if it is impossible for it to be true; that is, the sentence is inconsistent.

Logically indeterminate: A sentence is logically indeterminate (contingent) if and only if it is neither logically true nor logically false.

Logically true: A sentence is logically true if and only if it is impossible for it to be false; that is, the denial of the sentence is inconsistent.

Major premise: The first part is a general statement concerning a whole group.

Mean: The sum of the data in a data set, divided by the number of items in the set.

Median: The middle number of a data set when the data are in numerical order.

Median of a trapezoid: The line joining the midpoints of the nonparallel sides.

Member: Each individual item belonging to a set.

Midline: The line segment that joins the midpoints of two sides of a triangle; a median line of a trapezoid.

Minor premise: The second part is a specific statement which indicates that a certain individual is a member of that group.

Mode: The number or element in a data set that appears most often in the set.

Multiple: Any number that can be divided by another number with no remainder.

Multiplicative inverse: The reciprocal of a number.

Multiplicity: Number of times that $(x - r)$ is a factor of the equation.

Natural numbers: The set of counting numbers expressed as {1, 2, 3, ...}. This set is identical to the set of whole numbers, less the number zero. Natural numbers are not negative.

Negation: Given statement q, the negation is "not q."

Negation operation: If X is a sentence, then $\sim X$ represents the negation, the opposite, or the contradiction of X; $\sim$ is called the negation operation on sentences.

Negative integers: The set of integers starting with –1 and decreasing: {–1, –2, –3, …}.

Null set: A set with no members.

Number line: A line with equally spaced tick marks named by numbers.

Obtuse triangle: A triangle that has one obtuse angle (greater than 90 degrees).

Odd integers: The set of integers not divisible by 2: {… , –3, –1, 1, 3, 5, 7, …}.

Open interval: A continuous set of real numbers that does not contain its endpoints.

Order property of real numbers: If x and y are real numbers, then one and only one of the following statements is true: $x > y$, $x = y$, or $x < y$.

Parallel: Given two linear equations in x, y, their graphs are parallel lines if their slopes are equal.

Parallelogram: A quadrilateral whose opposite sides are parallel.

Permutation: An arrangement of a group, or set, of objects in a particular order.

Perpendicular: If the slopes of the graphs of two lines are negative reciprocals of each other, the lines are perpendicular to each other.

Perpendicular bisector: A line that bisects and is perpendicular to a side of a triangle.

Plane geometry: Refers to two-dimensional shapes (that is, shapes that can be drawn on a sheet of paper), such as triangles, parallelograms, trapezoids, and circles.

Point of tangency: A line that has one and only one point of intersection with a circle is called a tangent to that circle, and their common point.

Polygon: Any closed figure with straight line segments as sides.

Power: Indicates the number of times the base appears when multiplied by itself.

Premise: The compound statement "if a, then b" is called a conditional statement or an implication. "If a" is called the hypothesis or premise of the implication.

Prime numbers: The set of positive integers greater than 1 that are divisible only by 1 and themselves: {2, 3, 5, 7, 11, …}.

Product: The answer when two or more numbers are multiplied together.

Proper fractions: Numbers between –1 and 1; the numerator is less than the denominator.

Pythagorean Theorem: The lengths of the three sides of a right triangle are related by the formula $a^2 + b^2 = c^2$.

Quadratic equation: An equation in which the highest power of an unknown quantity is two.

Quadratic formula: If the quadratic equation does not have obvious factors, the roots of the equation can always be determined by the quadratic formula in terms of the coefficients a, b, and c:

$$x = \frac{-b \pm \sqrt{b^2 - 4ac}}{2a}.$$

Quadrilateral: Any polygon with four sides.

Radius: Fixed distance from the center of the circle to any point on the circle.

Range: The set of all the values of y in a relation; the difference between the greatest number in a list of numbers and the least.

Rational numbers: The set of all terminating and repeating decimals; is of the form, where p and q are integers, and q is not equal to zero.

Real number: Any positive or negative number; includes all integers and all rational and irrational numbers.

Real part: The real and imaginary parts of a complex number z are respectively the first and the second elements of the ordered pair.

Reciprocal: One of two numbers whose product is 1; two numbers are reciprocals of each other if their product equals 1.

Rectangle: A parallelogram with right angles.

Reflection: Movement of a figure to a new position by flipping it over a line.

Remainder theorem: If a is any constant and if the polynomial $p(x)$ is divided by $(x - a)$, the remainder is $p(a)$.

Response variable: Measures the outcomes that have been observed.

Rhombus: A parallelogram that has two adjacent sides that are equal.

Right triangle: A triangle that has a right angle.

Rotation: A function moves each point P to a new point P' so that $OP = OP'$.

Scalene triangle: A triangle that has no equal sides.

Scatter plots: A graph that shows data points on a coordinate grid.

Secant: A line that intersects a circle in two points.

Sense: Two inequalities are said to have the same sense if their signs of inequality point in the same direction.

Set: A collection of items.

Shape of a curve: A plot is usually classified as linear or nonlinear (curved).

Side of a triangle: A line segment whose endpoints are the vertices of two angles of the triangle.

Similar: A one-to-one correspondence between the vertices of a polygon such that all pairs of corresponding angles are congruent and the ratios of the measures of all pairs of corresponding sides are equal.

Simultaneous equations: Equations involving two or more unknowns that are to have the same values in each equation.

Skewed: The graph will either be skewed to the left, skewed to the right, or approximately normal. A skewed distribution of data has one of its tails longer than the other.

Slope-intercept form: The equations will be in the form $y = mx + b$, where m is the slope and b is the intercept on the y-axis.

Solution: A number that makes the equation true when it is substituted for the variable.

Square: A rhombus with a right angle.

Standard deviation: The standard deviation is a numerical value used to indicate how widely individuals in a group vary; the square root of variance.

Statement: A sentence that is either true or false, but not both.

Stemplot: Also called stem-and-leaf plot, can be used to display univariate data.

Strength of correlation: How tight or spread out the points of a scatter plot are.

Subset: Given two sets A and B, A is said to be a subset of B if every member of set A is also a member of set B.

Substitution: A method of solving systems of equations used to find an expression for the value of a variable from one equation and then substitute the expression for that variable in the other equation.

Subtraction: Process or skill of taking one number or amount away from another.

Syllogism: An arrangement of statements that would allow you to deduce the third one from the preceding two.

Tangent: A line that has one and only one point of intersection with a circle.

Transitive property of inequalities: If a, b, and c are real numbers with $a > b$, and $b > c$, then $a > c$. Also, if $a < b$ and $b < c$, then $a < c$.

Translation: A function will move each point of the function a specific number of units left or right, then up or down.

Trapezoid: For quadrilateral with two and only two parallel sides.

Triangle: A closed three-sided geometric figure.

Truth table: For sentence X, the exhaustive list of possible logical values of X.

Union of two sets: The union of two sets A and B, denoted $A \cup B$, is the set of all elements that are either in A or B or both.

Universal set: A set from which other sets draw their members.

Valid: The truth of the premises means that the conclusions must also be true.

Variability: The quality of being subject to variation.

Variance: The variance tells us how much variability exists in a distribution. It is the "average" of the squared differences between the data values and the mean.

Venn diagram: A visual way to show the relationships among or between sets that share something in common.

Vertex: For a parabola the point where it intersects the axis of symmetry; the highest or lowest point on a parabola.

Vertices: For a hyperbola the line through the foci that intersects the graph at two points; the points where the lines of the equations of two constraints intersect.

Whole numbers: The set of whole numbers can be expressed as {0, 1, 2, 3, ...}. This is the set of natural numbers and zero. Whole numbers are not negative.

Index

A

Absolute inequality, 63
Absolute value equations, 61–62
Addition, 50, 55, 56
Addition, subtraction, 65–66
Algebra
  absolute value equations, 61–62
  advanced theorems of, 73
  complex numbers, 65–66
  drill questions for, 74–77
  equations, 49–54
  exponents, 45–47
  fundamental theorem of, 73
  inequalities, 62–64
  logarithms, 48–49
  quadratic equations, 67–72
  simultaneous linear equations, 54–61
Altitude (height), 98, 104, 109
AND, 166
AND/OR, 167
Angle bisector, of triangle, 99
Antecedent, of implication, 160
Arc, of circle, 114
Area, of circle, 112
Area, of triangle, 98
Associative laws, of sets, 19
Associative property, 30

B

Base, 45, 108
Base angle, 109
Biconditional statement, 161
Bimodal set, 137
Bivariate data, 147

C

Center, of circle, 112
Central angle, 114
Central tendency, 136
Chord, of circle, 113
Circles
  circumscribed, 116

concentric, 116
congruent, 115
defined, 112
parts of, 112–115
Circumference, 112
CLEP examinations
  about, 5–6
  for candidates with disabilities, 7
  exam overview, 5
  for the military, 7
  online REA study center, 4–5
  preparing for, 3–4
  study plan for, 8
  test-taking tips, 8–10
Combinations, 127–128
Commutative laws, of sets, 19
Commutative property, 30
Complement laws, of sets, 18
Complex numbers, 65–66
Composite numbers, 35
Conclusion, of implication, 160
Conditional inequality, 62
Congruent circles, 115
Conjunction, of statements, 160
Conjunction operator, 166
Consecutive angles, 103
Consecutive integers, 34
Consequent, of implication, 160
Contrapositive, of statement, 160
Converse, of statement, 160
Counting rule, 125

D

Data analysis
  about, 142
  bar graphs for, 142–143
  histograms, 143–144
  line graphs, 144–145
  pie charts (circle graphs), 145–146
  stemplots, 146–147
De Morgan's laws, of sets, 19
Deduction, deductive reasoning, 164–165
Dependent equations, 58

Notes

Notes

Notes

Notes

Notes

Notes